The Chimney of the World

THE CHIMNEY OF THE WORLD

A History of Smoke Pollution in
Victorian and Edwardian Manchester

Stephen Mosley

The White  Horse Press

Contents

Abbreviations

MAPS   Manchester Association for the Prevention of Smoke
NVAA   Manchester and Salford Noxious Vapours Abatement Association
SAL   Smoke Abatement League

Illustrations

Figures

Tables

Acknowledgements

Many people have helped in making the writing of this book possible. I am particularly indebted to fellow historians John Walton, David E. Nye, Paolo Palladino, Thomas Rohkrämer and Mike Winstanley for their useful comments on early drafts of this work. I am also grateful to colleagues in other disciplines who found the time to discuss issues where I do not have specialist knowledge, and to read and advise on some sections of the manuscript. Professor T.A. Mansfield, Peter Lucas and Colin Wells of the University of Lancaster's Institute of Environmental and Biological Sciences, and Greg Myers of the University's Linguistics Department, provided me with important information and saved me from making several mistakes. Any errors that remain in the text are entirely my own responsibility.

I owe a large debt of gratitude to the librarians and archivists of the following institutions for providing generous assistance: the University of Lancaster Library; the University of Birmingham Library; the British Library; the Local Studies Unit of the Manchester Central Reference Library; the Salford Local History Library; the National Society for Clean Air, Brighton; and the Working Class Movement Library, Salford. I must also thank the Economic and Social Research Council for funding my research between the years 1994–97, and the University of Birmingham, Westhill, for allowing me study leave to finish the final draft of the manuscript. Special thanks must go to Alison and Andrew Johnson for their generous help in bringing the project to completion. My greatest debt is to my wife, Monika Büscher, who provided constant encouragement along the way. Without her invaluable support this book would not have been written.

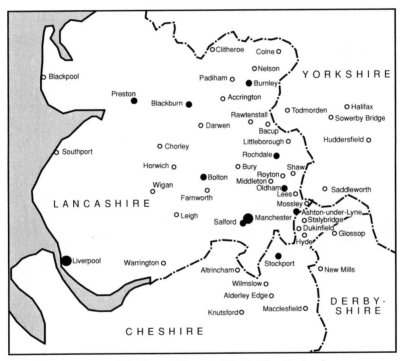

Map 1. The Cotton Towns of North-West England.
Adapted from Fowler, A. and T. Wyke, *The Barefoot Aristocrats*, 1987.

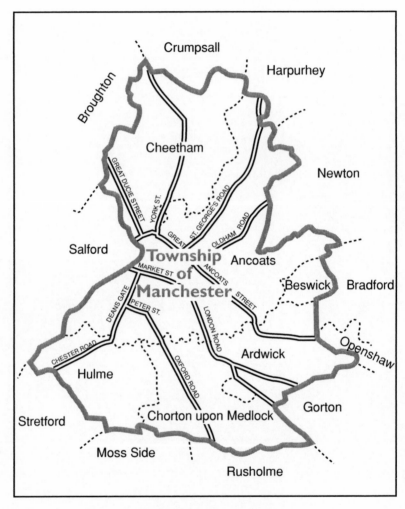

Map 2. Manchester in 1838.

Adapted from Redford, A., and Russell, I.S., *The History of Local Government in Manchester. Vol.II: Borough and City*, 1940.

The Chimney of the World

Introduction

Manchester, Air Pollution, and Urban Environmental History

In 1994, when I began researching this study, a national survey of air quality revealed that Manchester had the dirtiest air in Britain, with the possible exception of Belfast. Readings taken on the Manchester Town Hall roof on 22 December 1994, as 65,000 cars entered the city centre between the hours of 7am and 10am, showed high concentrations of both nitrogen dioxide and carbon monoxide.[1] Pollutants from vehicle exhaust fumes have been linked to an alarming rise in respiratory diseases and an increase in the incidence of a variety of cancers in urban areas.[2] Manchester's ascent to the top of the national air pollution league table sparked protests from environmental campaigners in the city, who swiftly formed an alliance called Fresh Air Now. Radical direct action groups, such as Reclaim the Streets, have sprung up all over the country in an attempt to halt new road-building schemes and bring cleaner air to Britain's cities.[3] The current level of public interest and involvement in actions to reduce urban air pollution is impressive but not wholly without precedent. Concerns about the air above cities have a long history, and are not, as some modern Green activists tend to think, simply a new product of late twentieth-century anxieties about global environmental degradation.[4] Moreover, just as today, nineteenth-century Manchester, the 'shock city' of the Industrial Revolution, also found itself in the spotlight regarding dirty air.

In February 1884 John Ruskin, considered by some modern commentators to be 'the first Green man in England', symbolically represented Manchester as the spiritual home of air pollution.[5] In his lectures on 'The Storm-Cloud of the Nineteenth Century' Ruskin described to his audience a 'terrific and horrible' thunderstorm observed from his home at Coniston Water in the Lake District, during the course of which the air became 'one loathsome mass of sultry and foul fog, like smoke' before finally blowing itself out, only to leave behind the sullen climatic conditions that he provocatively named 'Manchester devil's darkness'.[6] By the early 1880s, after a century of rapid urban and industrial growth, the name of Manchester had become synonymous with leaden skies, dirt and smoke. Ruskin was deeply worried about the effects that air pollution drifting in from the numerous towns and cities of south-east Lancashire might have upon the Lakes. And in choosing to highlight the smoky image of Manchester – the world's first real industrial city – as the concrete embodiment of his concerns, he had picked a fitting target.[7] Manchester, once fêted as 'the symbol of a new age', had come to epitomise the grimy, polluted industrial city: it was, in the words of one contemporary, 'the chimney of the world'.[8]

Today the shimmering haze of photochemical smogs has replaced the stifling gloom of sulphurous coal smoke as a major problem of urban life. Many scientists believe that rates of pollution emissions from an exponentially expanding human economy are a serious threat to the ability of the earth's natural cycles to regulate established climatic patterns and purify the air. Donella Meadows, Dennis Meadows and Jørgen Randers, for example, in their influential study *Beyond the Limits* declared, 'Human society is now using resources and producing wastes at rates that are not sustainable'.[9] More and more people, like Ruskin before them, are making the connection between local urban emissions and wider ranging pollution problems. Invisible, global phenomena such as the 'holes' in the ozone layer and the 'greenhouse effect' are now universally recognised as serious – and real – environmental threats. Against this backdrop it is not surprising to find that interest in the roots of air pollution problems is now increasing amongst historians. As concern for the environment has grown, so has the discipline of Environmental History, bringing a much-needed historical perspective to discussions about the ways in which humans interact with nature. This new field has started to dislodge the stubbornly recurring notion of a past 'Golden Age' of environmental harmony, when humans were more attuned to the earth and nature was pristine and untainted. It also provides a useful contextual framework for informed and critical debate about the implications of persistent pollution for future generations. Air pollution on a grand scale began with the Industrial Revolution, and, therefore, a study of the 'smoke nuisance' in Manchester, the spiritual home of both, may provide valuable insights for those concerned with finding solutions to today's urban environmental problems.

Manchester: setting the scene

By the 1850s Britain had become the 'workshop of the world': a sobriquet largely earned by the energy and enterprise of the Manchester region's industrial entrepreneurs and factory hands in supplying the world with cotton textiles. Some 90 per cent of Britain's cotton industry was concentrated in the smoky Manchester region (south-east Lancashire, north-west Derbyshire and north-east Cheshire) by 1835.[10] The employment of steam-powered technologies in both its spinning and weaving mills had made cotton Lancashire a centre of booming production and sustained economic growth 'the like of which the world had never seen before', as Tables 1 and 2 demonstrate.[11]

Date	Retained imports (million lbs)
1770–79	48
1780–89	155
1790–99	286
1800–09	594
1810–19	934
1820–29	1664
1830–39	3208
1840–49	4072

Table 1. Raw Cotton Consumption in Britain, 1770–1849.

Sources: Mitchell, B.R., *British Historical Statistics*, Cambridge University Press, Cambridge, 1988, pp.330–3; and Rose, M.B., (ed.), *The Lancashire Cotton Industry*, Lancashire County Books, Preston, 1996, p.7.

Date	£ (thousands)	% of total British exports
1784–86	766	6.0
1794–96	3,392	15.6
1804–06	15,871	42.3
1814–16	18,742	42.1
1824–26	16,879	47.8
1834–36	22,398	48.5
1844–46	25,835	44.2
1854–56	34,908	34.1

Table 2. Exports of British Cotton Goods 1784–1856.

Sources: Davis, R., *The Industrial Revolution and Overseas Trade*, Leicester University Press, Leicester, 1979, p.15; and Rose, M.B., (ed.), *The Lancashire Cotton Industry*, Lancashire County Books, Preston, 1996, p.9.

Burgeoning overseas demand for Lancashire's cotton goods in America, Europe, Africa and Asia had stimulated the industry's phenomenal growth, and the value of British exports in this commodity rose from 'practically nothing' (£46,000) in 1751 to almost £35 million a century later.[12] By the early decades

of the nineteenth century the Manchester region's cotton products constituted
well over 40 per cent of the value of all the nation's exports. The cotton textile
trade remained an important national industry throughout the Victorian and
Edwardian periods. Although its share of the value of British exports had fallen
to 25 per cent by 1913 it was still the nation's largest export industry.[13]
However, the Cotton Famine of the early 1860s, a major slump in trade
exacerbated by the American Civil War, exposed the myth that the industry was
essential for Britain's economic well being.[14] Even so, the Cotton Famine
brought mass unemployment and severe distress to Lancashire's factory towns,
as did other cyclical trade depressions, emphasising the crucial importance of the
industry at a regional level.

MANCHESTER, GETTING UP THE STEAM.

Illustration 1. 'Manchester, Getting Up the Steam.'

Source: *Builder* 1853.

The primary source of energy for Manchester's numerous steam-
powered mills – and the main source of heat in the homes of its many citizens
– was the abundant and inexpensive fossil fuel of the Lancashire coalfield. As coal
consumption increased, dense emissions of black, sulphurous smoke in Manchester gradually became a serious problem, blocking out the sun, destroying
vegetation, corroding buildings and damaging the health of city dwellers. In
1872 the term 'acid rain' was neologised by the Victorian scientist Robert Angus
Smith to describe some of the deleterious environmental consequences of air
pollution in and around Manchester.[15] Most contemporaries, however, commonly referred to the 'smoke nuisance' when discussing the myriad and far-

reaching effects of the city's air pollution problem. By the early 1840s, as vegetation was all but banished from the city centre, and mortality statistics began to reveal the extent of health problems caused by polluted air, an anti-smoke movement sprang into existence at Manchester. But while Manchester and other major British cities were fairly quick to tackle the pressing problem of a clean water supply, and the removal of other 'nuisances' such as refuse and human wastes, campaigners were still lobbying for the abatement of coal smoke well over a century later.[16]

Historians have often suggested that indifference on the part of the general public was the critical weakness that hindered the smoke abatement movement. Carlos Flick neatly sums up this widely held notion: 'the movement for smoke abatement, like the steam engines themselves, could not work without sufficient pressure behind it'.[17] 'In the absence of popular alarm and indignation', Flick continues, the problem of curbing smoke pollution 'seemed insuperable'.[18] Yet there is ample evidence to show that large numbers of people, including a good many members of the working classes, were highly indignant where the destructive effects of the 'smoke nuisance' were concerned. For example, millions of Victorians were engaged in an arduous daily battle with smoke and soot as they endeavoured, against the odds, to keep their houses and belongings clean. But popular indignation did not translate into a groundswell of support for the anti-smoke activists. The lack of widespread public alarm about a severe pollution problem that affected almost every aspect of day-to-day life has yet to be accounted for satisfactorily. Many historians have pointed to the fact that the Victorians commonly linked smoke pollution to industrial progress, employment and prosperity, but to date there has been little sustained engagement with this important issue. Nor has the other side of the coin, smoke pollution's negative association with waste and inefficiency, been addressed in sufficient historical detail.[19] In *The Chimney of the World* I aim to fill in these gaps, and to explain and analyse the complex reasons why the 'smoke nuisance' proved to be such an intractable problem. But as this study builds upon some outstanding work already accomplished by scholars in the field, it is first necessary to comment briefly on the historiography of the relatively new discipline of urban environmental history.

The historiographical setting

In 1984 the leading environmental historian Donald Worster declared, 'There is little history in the study of nature, and there is little nature in the study of history.'[20] And until a short time ago it is fair to say that British historians had largely neglected environmental issues. Initiatives to develop environmental history in Britain really started to take off following an international workshop on European Environmental History held at Bad Homburg, Germany, in 1988. The resulting publication, *The Silent Countdown,* edited by Peter Brimblecombe and Christian Pfister, provides a useful comparative discussion of the origins and environmental impact of urban-industrial pollution in England, Germany and the Netherlands. This upward trend has continued with the launch of the journal *Environment and History* in 1995, the establishment in 1998 of the Association for the Study of Literature and the Environment (UK), which incorporates a History and Culture strand, and the foundation of the European Society for Environmental History in 1999. While British environmental history is still maturing, on the other side of the Atlantic the field has come of age. The American Society for Environmental History has been exploring the intricacies of society-environment relationships for over twenty years.[21] Moreover, in France the work of the *Annales* school, and in particular Fernand Braudel's emphasis upon *la longue durée* in history, has also been successful in focusing attention on *histoire écologique* for several decades.[22] Since Worster's pronouncement, such has been the upsurge of interest in the topic of environmental history that a comprehensive review of the literature is unfeasible here. The vigorous growth of this new field of study both in Britain and overseas can be traced, and is attested to, by the numerous review essays concerning recent developments in environmental history.[23] Since this is not the place to undertake a detailed survey, what I propose to do instead is to focus on the emergence of urban pollution studies – the body of literature that inspired this book – as a key component of environmental history.

Up until the early 1990s the major preoccupation of environmental historians had been the study of 'the wild and the rural'. Indeed, Worster deliberately placed the city outside the purview of environmental history as 'The built environment is wholly expressive of culture'.[24] While he conceded that this exclusion 'may seem especially arbitrary, and to an extent it is', Worster went on to argue that, 'The distinction ... is worth keeping, for it reminds us that there are different forces at work in the world and not all of them emanate from humans'.[25] However, the majority of the world's six billion population now live in swelling towns and cities, and not on farms or in villages.[26] Moreover, today's growing environmental crises are in no small part the results of the activities and lifestyle patterns of Western urban dwellers. The time is ripe, therefore, to include the urban environment and its culture as one of the main themes of environmental history, and the rural focus of the discipline has begun to change.

Environmental historians such as William Cronon, Martin Melosi, Christine Meisner Rosen and Joel Tarr are presently directing attention away from wilderness regions and agriculture to explore the problems of the urban and industrial past.[27] In 1993, for example, Melosi challenged Worster's agroecological paradigm for environmental history when he argued:

... how can we justify as part of the *main* theme of environmental history the study of human intrusion in the natural world in the form of farming, and not in the building of a town or city? ... Excluding cities from the *main* theme of environmental history seems to be more of a rhetorical device than a well-crafted definition.[28]

The following year Rosen and Tarr called for more research into the different ways that natural and cultural forces shape modern cities and modern urban life, evolving together in 'dialectical interdependence'.[29] In this book I have endeavoured to respond by examining the influence and interaction of factors such as climate, population growth, technologies, cultural values and everyday activities in shaping the ways in which both people and place slowly adapted to the heavily polluted atmospheric conditions. However, the study of the urban environment, as Worster rightly points out, is well advanced in other fields.[30] The debt that my research owes to existing work undertaken within the mainstream of urban history, economic and social history, medical history, and the history of science and technology is acknowledged in both the footnotes and bibliography.

But in mainstream approaches to the Victorian city the natural environment is not usually an active historical agent. Nature acts for the most part as the quiescent stage upon which the affairs of humans are played out. In part one of this study I have attempted to show that the non-human surroundings themselves were much more than a passive backdrop to the unfolding historical narrative. The very essence of the discipline of environmental history is to reveal how, in conjunction with human actions, 'nonhuman entities become the coactors and codeterminants' of historical processes.[31] The first part of the book looks at the effects of smoke pollution on air quality, human health, and the natural and built environment, clearly showing that society-nature interactions are indeed a two-way street. While part one mainly deals with the material substance of environmental degradation caused by air pollution, reconstructing what it might have been like to live and work under Victorian Manchester's dense pall of smoke, it does not make any grand claim to objectivity. The shadowy cityscape that emerges cannot perfectly reflect the 'realities' of how people and place evolved together. Manchester as the 'chimney of the world' is a historical construct built up using a wide range of source materials, from working-class autobiographies to scientific investigations, all of which present different problems of selection and interpretation for the historian. Nonetheless, as David E. Nye has noted,

If historians cannot pretend to a complete objectivity, they can establish a bedrock of facts that set limits to what is true and what is false. Matters of chronology, measurement, kinship, and census-taking may be open to debate, but usually they can be established. Whatever the problems of representing such facts in language as part of an argument, historians work not with a 'rubble of signifiers' but a brickyard of facts.[32]

Nye's particular point is that although we are faced with many difficult 'problems of representation ... we are not writing fiction'.[33] But how individuals and communities understand issues relating to environmental change is not only a matter of the establishment of 'natural facts': it is also a question that turns on other representations of 'reality'.

Part two of this study embraces the recent 'linguistic turn' in historical studies. Post-modern critical theory has focused attention on the way in which the forces of social construction play a major role in forming human understandings of the natural world.[34] 'Nature' is in no small part a human idea and different people and cultural groups perceive environmental change in different ways at different times, dependent upon both context and situation.[35] Yet, as David Demeritt has observed, social constructivist ideas about representations of nature have only recently begun to move beyond the programmatic to influence environmental historians' work.[36] As many environmental historians have their roots firmly based in the natural sciences, more than a few have reacted with 'great fear and anxiety' to the notion that nature and the past are things we construct.[37] One of the few environmental historians to utilise the linguistic turn is William Cronon in his analysis of the history of the Great Plains, *A Place for Stories: Nature, History and Narrative.*[38] Significantly, however, Cronon refuses to adopt a strong relativist position, where nature becomes solely a construct of language and the imagination, insisting that 'given our faith that the natural world ultimately transcends our narrative power, our stories must make ecological sense'.[39] Indeed, in relating the narratives of the dust storms of the 1930s Cronon makes nature a co-author: 'You can't put dust in the air – or tell stories about putting dust in the air – if the dust isn't there.'[40] I have followed Cronon's example in trying to accommodate the lessons of social constructivist theory 'without giving in to relativism'.[41] The second part of this study is concerned with recovering and analysing the narratives that contemporaries employed to make sense of their polluted surroundings, and, more importantly, with showing that these stories about smoke had real consequences in the natural world.

The third and concluding part addresses contemporary responses in relation to tackling atmospheric pollution. The political and legislative history of the struggle for clean air in Britain, at both national and local levels, is well-charted terrain and still proliferating.[42] The politics of air pollution in Victorian Manchester and the laws relating to its control are important factors regarding the 'smoke question', and as such they are discussed in some detail. However, by scrutinising these issues in the broader context of the 'nature' of, and 'stories' about smoke pollution *The Chimney of the World* offers a richer insight into the ways in which nineteenth-century legal and administrative responses to environmental change evolved. The most influential historical studies of Britain's 'smoke nuisance' to date, those by Eric Ashby and Mary Anderson, and Anthony Wohl, accept that 'slow, piecemeal, persuasive legislation' was needed to reduce air pollution.[43] However, on the basis of my own reading of the primary sources I will argue that the overcautious authorities could have done more to prevent smoke. This argument is partly based on an examination of the smoke abatement technologies available to contemporaries: a path previously travelled by the aforementioned Carlos Flick. Flick is very informative about the numerous devices on the market and their design, while he also outlines the social and economic factors that limited their wider application. I take Flick's work a step further by investigating in more depth the social shaping of smoke prevention technologies, and also by examining the inapt ways in which they were often utilised by contemporaries.[44] I attempt to show that, on the whole, even when judged by their own criteria, the Victorians were not using the 'best practicable means' where smoke abatement was concerned. This book also complements work undertaken on river pollution, water supply and waste disposal in Victorian Britain, most notably Dale Porter's and Bill Luckin's innovative studies of the River Thames.[45] However, as space is at a premium, no sustained parallels between the treatment of air, water and land pollution in the nineteenth and twentieth centuries can be developed here.

A note on methodology

To better understand the complexities of the 'smoke nuisance' in Victorian Manchester I have employed an analytical framework that is loosely based on Rosen and Tarr's fourfold agenda for the study of urban environmental history: 1) the analysis of the impact of cities on the environment over time; 2) the analysis of the impact of the natural environment on cities; 3) the study of societal response to environmental problems; and 4) the examination of the built environment and its role and place in human life.[46] The ecological, economic, technological, social and cultural dimensions of urban air pollution are woven together here to produce a synthetic form of environmental history that provides a wide scope for inquiry and explanation. By directing attention to both the transformation of the physical environment and the ways in which language and cultural symbols were used to rationalise, naturalise or criticise the changes wrought by air pollution, a complex web of interconnections between nature and culture is revealed. What is offered, then, is a tentative model that not only aims to show the environmental consequences of urban air pollution, but that also explores the experiences, values and beliefs of the men and women who lived under a permanent canopy of smoke. I am especially concerned to discover why it was that a whole community – from Manchester's ordinary factory hands to its leading merchants and industrialists – allowed smoke pollution to contaminate the atmosphere of south-east Lancashire for well over a century. To do this I borrow ideas from social anthropology and phenomenological sociology that help to clarify the intricate connections between nature, culture and pollution beliefs.

A major problem in environmental history, as Christopher Hamlin has observed, 'is understanding why people came to see a problem and to take action. We need to know who cared, why they cared, and we must try to explain why an objectionable condition became a public issue at a particular time'.[47] The work of the social anthropologist Mary Douglas, a leading authority on the subject of pollution, provides a useful conceptual guide when exploring these kinds of issues. In her seminal work *Purity and Danger* Douglas argues that,

> If we can abstract pathogenicity and hygiene from our notion of dirt, we are left with the old definition of dirt as matter out of place. This is a very suggestive approach. It implies two conditions: a set of ordered relations and a contravention of that order. Dirt, then, is never a unique, isolated event. Where there is dirt there is system.[48]

Douglas's emphasis on the concept of dirt as 'matter out of place' compels us to think long and hard about when and why a community might – or might not – condemn coal smoke as pollution. And if close attention must be paid to explaining the contexts in which smoke might be deemed polluting, the accent she places on order and disorder foregrounds the question of how people use ideas about pollution as support for a certain type of society. Douglas's notion that dirt implies system brings into sharp relief the way that the pollution beliefs of a community are related to a specific, ordered way of life. Each culture, she argues, selects pollution dangers in accordance with what it believes to be 'normal or natural ... Common values lead to common fears (and, by implication, to a common agreement not to fear other things)'.[49] The views of Mary Douglas enable us to begin to make sense of why smoke pollution was tolerated for so long in Victorian Manchester, while other environmental dangers were not.

If social anthropology is highly relevant where inquiries into pollution behaviour are concerned, so, too, is phenomenological sociology. The phenomenological perspective of the Austrian philosopher Alfred Schutz concerning what we know about the life-world, and how we come to know it, is particularly useful for the investigation at hand.[50] Schutz's work opens up a vista onto the mechanisms through which knowledge about a system of pollution beliefs could be shared within an urban-industrial culture. The world we live in precedes us. People make sense of their surroundings as members of an intersubjective 'world known in common'.[51] 'Intersubjective' draws attention to the fact that experiences of the smoky urban-industrial environment, stories about it, and people's perceptions of it, were shared in the recurring rhythms of everyday life. Common attitudes to the 'smoke nuisance' were built up by dint of mutual participation in the daily round of mundane social activity, and active engagement with the polluted environment, over the course of generations. Some insight into what kinds of interpretations and perceptions pervaded this system of knowledge in Victorian Manchester can be gained through an examination of a wide range of primary source materials. Contemporary poems and songs, housekeeping manuals, cartoons, postcards and the vernacular literature of the Lancashire working-classes provide a rich and so far largely unexplored entry-point for the study of nineteenth-century pollution behaviour and beliefs.

Furthermore, wherever possible, contemporaries will be allowed to speak for themselves. This is to ensure that their concerns and views about the 'smoke nuisance' are accurately expressed and to demonstrate that there are powerful and compelling continuities – as well as some startling discontinuities – between past and present attitudes toward air pollution problems. Inevitably, gaps do appear in the source materials and they were supplemented where possible by using evidence gleaned from other large towns and cities. By the late-nineteenth century cooperative links were being forged between anti-pollution activists in Manchester, Glasgow, Leeds, London, Sheffield and elsewhere, and these links were being extended as far afield as Germany and America by the early twentieth century.[52] While each place had its own unique air pollution problems, smoke did have similar effects in many manufacturing centres both in Britain and overseas. And, as the reformers themselves shared both ideas and information about the 'smoke nuisance', I believe that it is justifiable to fill in the gaps in this way in order to provide a fuller picture. However, it is not a desire to identify and pay homage to 'heroic' precursors of the present Green movement that lies at the heart of this study. It is a wish to provide a broader and more critical historical study of urban air pollution in Britain than has hitherto been produced.[53]

With regard to organisation, I am aware that by dividing this book into three distinct segments on 'nature', 'stories' and 'responses' the old fault lines between culture and nature are clearly visible. Nonetheless, the three main parts of this study – despite their analytical separation and linear sequence – should be seen as mapping out different facets of an interrelated whole. They should be viewed in parallel, each part complementing the other, rather than as the vehicles of an argument that progresses directly towards its conclusion. Taken together, the three parts of this book provide a fresh environmental perspective on Victorian Manchester's commercial 'success story': an eco-historical perspective that is highly relevant for a better understanding of present air pollution dilemmas in a world obsessed by economic growth.

Part One

The Nature of Smoke

South-east Lancashire witnessed an unprecedented burst of urban-industrial activity during the early decades of the nineteenth century, with Manchester the focal point of this explosive expansion. Smoke pollution subsequently became a major problem as emissions from growing numbers of coal-fuelled factories and domestic hearths permanently filled the air. In this part of the study I analyse how Manchester's natural and built environments evolved in dialectical interconnection with anthropogenic air pollution.[1] Moreover, as well as looking at the urban ecosystem in which this environmental dilemma developed, the urban culture that generated the smoke will also be considered. Without taking into account the socially transmitted behaviour patterns concerning the production of air pollution that were characteristic of the northern industrial towns of the period, the intricacies of the 'smoke nuisance' cannot begin to be fully appreciated. The limited scientific data of the period (and more recent scientific studies) can help to quantify the extent of the city's air pollution problem and build up a picture of what it might have been like to live in smoky 'Cottonopolis'. But here, by including non-scientific lay evidence alongside data provided by scientific and professional 'experts', a broader and richer environmental understanding of the past begins to come into focus. By asking the largely unexplored question 'How did ordinary people connect their own lives to the environment?' I hope to shed some light on how the public experienced smoke pollution and how they coped with it.[2] For people's perceptions and values regarding environmental conditions – especially about the place in which they live – are formed in no small part by their active engagement with the physical world.

The historical geographer Yi-Fu Tuan stressed the importance of how humans actually experience living in their different cultural worlds to the development of their environmental perceptions, attitudes and values over two decades ago.[3] And the experiential message that 'the world we perceive comes into being as we act in it' has now been taken up in earnest by scholars in the social sciences.[4] The cultural anthropologist Kay Milton has recently emphasised that human-environment relations are formed not only through socially constructed knowledge (including scientific analysis), but also by sensory perceptions of the world.[5] Arnold Berleant, a leading proponent of the emerging discipline of environmental aesthetics, speaks of the environment as 'a seamless unity of organism, perception, and place', and asserts, 'Environment is not the construction of a perceiver or the geographical character of a place, or even the sum of these. It is their original unity in active experience.' [6] People live *in* the environments that embrace them, and are not in some mysterious way able to hold themselves aloof from them. Knowledge of the city's physical surroundings and the ways in which Manchester's inhabitants lived out their everyday lives must, therefore, be brought together if the full complexity of the evolving 'smoke nuisance' is ever to be properly understood.

Coal, steam and people

By the early years of the twentieth century many of Britain's towns and cities had enjoyed a considerable measure of success in cleaning up environmental 'nuisances' relating to land and water, while, in stark contrast, a solution to the problem of reducing smoke pollution in urban areas remained elusive. 'Insanitary conveniences, defective drains, foul methods of sewage disposal, polluted streams, are all being steadily improved', observed the *Builder* in 1899, before continuing, 'But while other nuisances are being gradually abated, the smoke nuisance increases year by year'.[7] The sense of achievement that many contemporaries undoubtedly experienced as damage caused to the health and property of urban dwellers gradually abated with every new advancement in town planning and water and sanitation systems was tempered by the notion that smoke pollution in Britain was going from bad to worse. However, the idea that the 'smoke nuisance' in Britain's cities was a rapidly proliferating problem, due to sustained urban and industrial growth, was already being hotly debated at the beginning of the nineteenth century. The 1819 Select Committee on steam engines and furnaces was told that the nuisances caused by factory smoke were 'daily increasing' in Britain's manufacturing centres, with Birmingham and Manchester already thought to be smokier than London.[8] In 1843 the Select Committee on Smoke Prevention concluded that 'the evils arising from smoke are severely felt in all populous places, and are likely to increase in proportion as wealth and the use of machinery cause a greater extension of furnaces and steam-engines'.[9] The Reverend John Molesworth's evidence to the same committee protested: 'In Manchester there are nearly 500 chimneys discharging masses of the densest smoke; the nuisance has risen to an intolerable pitch, and is annually increasing'.[10] Almost half a century later a correspondent to the *Manchester Guardian* thundered,

Manchester is doomed to perpetual smoke! Our city, which ought to be one of the finest in the kingdom, is to be entirely at the mercy of the smoke and noxious vapour producer, with the result that the atmosphere is poisonous, mortality at its highest rate, and altogether produces a most depressing effect upon the spirits of those unfortunate inhabitants who are compelled to live within its boundaries … Manchester has an unenviable reputation. It is time to roll away the reproach.[11]

By the late 1880s Manchester's dismal image as one of the most smoke-polluted cities in Britain was certainly assured. But the numerous complaints of its citizens about smoke over the years give few real clues as to whether atmospheric conditions had deteriorated in the city since the beginning of the century. The perceived upward trend in the growth of the 'smoke nuisance' remained a serious cause for concern in Manchester (and the rest of industrialising Britain) throughout the nineteenth century. But was the problem getting worse as many contemporaries feared?

Smoke pollution was not a new phenomenon. It was a nuisance in large cities such as London long before the Industrial Revolution of the eighteenth and nineteenth centuries.[12] However, until the onset of the Industrial Revolution the demand for coal in south east Lancashire was fairly modest.[13] The Lancashire coalfield had produced fuel for industrial use since the thirteenth century.[14] But the lack of large towns, the difficulties of transporting coal, and the abundance of cheap or free alternative fuel for domestic use in the form of wood and turf meant that before 1700 the Lancashire coalfield was 'broad in area but small in output'.[15] In the late seventeenth century the total output of Lancashire coal was probably only a little over 50,000 tons per annum, making it one of 'the least productive and least dynamic of British coalfields'.[16]

But as the availability of cheap wood supplies declined coal became an essential commodity for many of Manchester's citizens. As early as 1615 the town's domestic hearths perhaps consumed as much as 10,000 tons of Lancashire's annual coal production, chiefly supplied from the nearby manor of Bradford.[17] By 1764 the construction of the third Duke of Bridgewater's canal had both increased the town's coal supplies and reduced prices for its poorer inhabitants. Coal imported from the Duke's Worsley mine into Manchester rose from 17,000 tons in 1765 to 52,000 in 1782, with the figure reaching 102,000 tons in 1800.[18] By the early nineteenth century the expansion of the region's towns and cities, a much improved transport infrastructure, and the burgeoning energy requirements of its cotton industries saw the Lancashire coalfield as a whole enjoying dramatic rates of growth in coal production. The landlocked coalfield's distance from the sea meant, however, that little of

Lancashire's coal was exported abroad. Stiff transport costs placed it at a considerable disadvantage compared with coalfields in Cumberland, Scotland and South Wales. Moreover, coal mined in Yorkshire and Staffordshire was finding its way into Lancashire by this time, with Manchester becoming the main market for Staffordshire coal.[19] In 1836 James Wheeler described Manchester's favourable position 'in the very heart of the great coal-field of England':

> ... Manchester has, in her own immediate vicinity, a copious supply of coal from the mines of Pendleton, Pendlebury, Worsley, Ashton, Duckenfield, Oldham, Rochdale, Middleton, Radcliffe, Tonge, Great and Little Lever, Darcy Lever, Hulton, &c; and travelling beyond this circle of about ten miles, there is the second or Wigan coal district, embracing the districts of Hindley, Abram, Leigh, &c ... The weekly consumption of the town and neighbourhood is estimated at about 26,000 tons ...[20]

There are no reliable series of output statistics for coal production in Britain before 1850. Nor are there any continuous statistical series for the consumption of coal in Manchester for the nineteenth century.[21] The importance of the evidence presented below in Table 3, despite the fragmented and uncertain character of its enumeration, is that it clearly shows both the inexorable increase year on year of coal production in Lancashire and Britain and the voracious consumption of fossil fuel in Manchester.

Output	Lancashire (m. tons)	Great Britain (m. tons)	Consumption	Manchester (m. tons)
1811–15	1.6	17.4	1800	0.1
1831–35	4.5	31.6	1836	0.9
1854	9.9	64.7	1854	2.0
1875	21.0	133.3	1876	3.0
1913	28.1	287.4	1913	n.a.

Table 3. Coal Output in Lancashire and Great Britain and Coal Consumption in Manchester, 1800–1913.

Sources: Lawton, R., and Pooley, C.G., *Britain 1740–1950: An Historical Geography*, Edward Arnold, London, 1992, p.69; Flinn, M.W., *The History of the British Coal Industry, Volume 2, 1700–1830: The Industrial Revolution*, Clarendon Press, Oxford, 1984, p.231; *Manchester As It Is*, Love & Barton, Manchester, 1839, p.26; Spence, P., *Coal, Smoke, and Sewage, Scientifically and Practically Considered*, Cave & Sever, Manchester, 1857, pp.7–8; Smith, R.A., 'What Amendments are required in the Legislation necessary to prevent the Evils arising from Noxious Vapours and Smoke?', *Transactions of the National Association for the Promotion of Social Science*, 1876, p.518.

The consumption of coal in Lancashire and Britain as a whole grew steadily throughout the eighteenth century, and then increased dramatically in the first fifty years of the nineteenth century: principally due to coal's 'new industrial uses – iron-making, gas supply, and the many and varied industrial uses of steam-power'.[22] It is difficult to exaggerate the importance of the role that the cotton industry played in encouraging the use of steam power in Manchester. For it was in 'Cottonopolis' that, in the words of Matthew Boulton, factory owners first went 'Steam Mill Mad', with Lancashire going on to become the leading market for Boulton and Watt steam engines.[23] Historians continue to debate the size of cotton factories in the city and whether or not Manchester was primarily a commercial or a manufacturing centre.[24] Important as these considerations are, the crucial point to remember here is that the cotton trade, whether embodied in the shape of a large firm or small, or in the form of a warehouse or factory, was a major source of smoke in the city during the period. For example, Charles Estcourt, City Analyst for Manchester, directed attention to the 'immense volumes of black smoke' which issued from 'the numberless small chimneys of warehouses which pack their own goods'.[25] Indeed, by the late nineteenth century there were few substantial industrial and commercial cotton concerns in Manchester and Salford that did not use a steam engine of some kind.

In the early 1780s Manchester's first cotton mill to utilise a coal-fuelled steam engine was established by Richard Arkwright, in partnership with Simpson and Whittenbury, at Shudehill.[26] After this date the centralisation of Lancashire's mushrooming cotton mills in towns, as adequate access to water power declined, was increasingly dependent on the use of steam. John Walton has observed that by the 1820s the 'balance of power output had probably tilted from water to steam' in cotton Lancashire.[27] Nationally, water wheels accounted for only 14 per cent of all horsepower generated in the cotton industry by 1850.[28] Cotton spinning in Lancashire was essential to the fortunes of steam engine makers. For, as A.E. Musson has emphasised, 'It is not generally realised … how long the predominance of the cotton industry in the employment of steam power persisted, and how slowly by comparison steam engines were introduced into other industries'.[29] While acknowledging that our statistical picture of the growth of Britain's manufacturing industry is far from complete, Musson estimates that as late as 1870 considerably more than 50 per cent of total manufacturing horsepower was used in textile production, with cotton factories alone utilising well over 30 per cent.[30] The research undertaken by many historians over the years has shown the dramatic multiplication of cotton mills in Manchester, which numbered 26 in 1802, 66 by 1821, and 182 in 1838. Taking Manchester and Salford together, as contemporaries often did, in 1816 there were 86 steam-powered spinning factories in the twin cities. In 1870 textile works of all kinds numbered 1,333 in Manchester and 296 in Salford, providing employment for some 43,615 people and 14,377 people respectively.[31] However, cotton was not the sole basis of Manchester's and Salford's manufacturing economies. Both cities contained a diverse range of industries from iron, rubber, and chemical works through to food processing plants and breweries, most of which burned bituminous coal to obtain steam-power. In the early 1780s the tall chimney of Arkwright's mill had stood alone. By the early 1840s some 500 industrial chimneys had sprouted in Manchester. This figure had more than doubled to around 1,200 chimneys by 1898 with neighbouring Salford adding a further 760 smokestacks of its own: more by far than almost any other town or city in the country.[32]

In addition, there were thousands of domestic chimneys in Manchester that also polluted the atmosphere of the city. Although by 1830 industry had outstripped the domestic hearth as the main consumer of coal in Britain, the sheer scale of population growth in Manchester and Salford contributed substantially to the 'smoke nuisance'.[33] Bituminous coal was the main source of heat and power in the homes of Britain's new urban masses, and, as the data in Table 4 confirms, population growth in the twin cities was considerable.

	1801	1851	1861	1871	1881
Manchester	75,281	303,382	338,722	351,189	341,414
Salford	18,179	85,108	102,449	124,801	176,235

Table 4. Population Growth of Manchester and Salford, 1801–81 (Municipal Boundaries).

Note: Commercial expansion in the heart of Manchester saw population figures decline for the central area of the city after 1871.

Of coal consumption per capita in 1830s Manchester, Wheeler noted, 'It has been calculated, that in London each individual consumes a ton of coal in the year. In these districts the consumption will, no doubt, be greater.'[34] In March 1866 Sir Robert Peel, son of the former Prime Minister, addressed the following rhetorical question to the members of the House of Commons:

> What is the coal consumption of Manchester? … At this moment the population of London, I believe, is about 3,000,000, and the annual consumption of coal amounts to about 5,300,000 tons; but in Manchester, with a population of certainly not more than 380,000, the coal consumption is estimated at 2,000,000 tons per annum, within a radius of three miles from the Exchange.[35]

Using Peel's figures London's inhabitants were each consuming 1.77 tons of coal per annum in 1866, in comparison with a prodigious 5.26 tons per capita in Manchester. If we take Robert Angus Smith's estimate of three million tons of coal consumed in the city by the 1870s and set this total against the 475,990 population of the two municipal boroughs of Manchester and Salford, we see a dramatic increase to 6.30 tons per capita. Roy Church has estimated that coal consumption in Britain as a whole saw a threefold increase between 1830 and 1913, when it slightly exceeded 4 tons per capita.[36] Coal exports also rose from less than 2 per cent of British output in 1830 to about 10 percent in 1870, increasing to some 27 per cent in 1913.[37] Most of Britain's expanding coal production was, then, still utilised at home. And Britain continued to consume more coal per capita than any other country in the world until 1905, when the United States recorded a higher figure. Chicago, one of the smokiest cities in the United States, consumed some 6.91 tons of coal per capita in 1910.[38]

Taken as a whole, an analysis of the foregoing statistical data regarding coal production and coal consumption in the Manchester region shows a striking increase in the output and utilisation of coal in order to maintain and promote urban and industrial growth over the period. Cotton manufacturing was initially of crucial importance in bringing the new fuel technology into the city, with the large-scale application of steam power to various other branches of industry being introduced more slowly during the second half of the nineteenth century. By the end of the nineteenth century most businessmen in the twin cities employed coal-fuelled steam engines in their warehouses, textile factories, printing, iron, chemical, or engineering works: there were even steam-powered laundries.[39] In addition, most of Manchester's many thousands of householders had come to prefer to burn coal to obtain domestic heat and energy. Thus, as Manchester's businessmen and householders rarely used coal efficiently, the statistical evidence strongly suggests that the smoke cloud was gradually deepening throughout the period. First designated as a serious environmental problem by contemporaries in the city during the 1840s, by the time it was 're-discovered' by a second generation of reformers in the 1870s the 'smoke nuisance' had undoubtedly intensified. Even so, it must be emphasised that this was probably experienced as a cumulative escalation of the nuisance and not as a sudden, spectacular surge. Nonetheless, in the nineteenth century Manchester was one of the smokiest cities in the world.

Sensing smoke

Unlike many of today's unseen environmental threats – such as climate change or the thinning of the ozone layer – the advancing 'smoke nuisance' did not elude the sensory perceptions of contemporaries. Coal smoke characterised the nineteenth-century urban atmosphere and affected the lives of all city dwellers, rich and poor alike. At the inaugural meeting of the Manchester Association for the Prevention of Smoke on 26 May 1842, the Reverend John Molesworth, the association's chairman, vehemently denounced a nuisance that 'polluted our garments and persons', and which all the town's inhabitants 'saw, tasted, and felt'.[40] But, despite the tangible nature of the problem, most contemporaries gave very little thought as to the composition of the coal smoke that surrounded them.

Coal is mainly composed of carbon, with smaller amounts of oxygen, hydrogen, nitrogen and sulphur also being present. The majority of coals contain between 85–90 per cent carbon.[41] The complete combustion of coal, the products of which are water vapour, carbon dioxide, and ash (non-combustible inorganic residue), does not produce any smoke. Small particles of carbonaceous matter – smoke and soot – are produced only when the coal is not completely burned. The high density of smoke and soot in Manchester's air was the result of the inefficient combustion of bituminous coal in the city's domestic grates and industrial furnaces. In addition, the smoke and soot particles absorbed oily, tarry hydrocarbons, most often produced by lower temperature fires in domestic grates, which, when combined together, formed the sticky 'blacks' or 'smuts' that contemporaries complained so much about. The formation of smoke was also accompanied by high concentrations of acid in the city's atmosphere, which the viscous smoke and soot particles also merged with in foggy weather. The sulphur content of coal rapidly oxidises during combustion to form the volatile invisible gas sulphur dioxide, which in turn readily combines with moisture in the air to form sulphurous acid, H_2SO_3, and the highly corrosive compound H_2SO_4, sulphuric acid or 'oil of vitriol'. The sulphur content of coal normally varies between 0.5 and 4.0 per cent, with the average amount being around 1.3 per cent.[42] Lancashire's coal, however, had a fairly high sulphur content of around 1.4 to 2.9 per cent, and contemporaries soon grew familiar with its very unpleasant and very palpable properties.[43] Smoke could be readily perceived by four of the five senses: one could see it, smell it, touch it, and it could be tasted. During the course of the nineteenth century sulphurous, viscid clouds of smoke gradually engulfed Britain's industrial cities and their inhabitants, and no one could be insensible as to smoke's effects.

In 1819 W. Frend wrote, 'The Smoke of London, first viewed from a distance, affords a sight which strikes a foreigner with astonishment.'[44] Peter Brimblecombe has compiled a large number of contemporary reactions to London's smoke and fogs, both of foreign visitors and native inhabitants, in his book *The Big Smoke*.[45] Contemporary impressions of Manchester's polluted atmosphere are also well chronicled, so here I will present only a select few accounts of the visual experience of the city's 'eternal smoke-cloud'.[46] Up until the end of the eighteenth century Manchester and Salford remained predominantly verdant and countrified towns, with Manchester even boasting its own pack of hounds.[47] Joseph Aston's *A Picture of Manchester* of 1816 still describes nearby Ardwick Green as a pleasant rustic suburb of Manchester, and opines that the residents of Salford Crescent would always be assured a view of 'rich rural scenery' from their front windows.[48] In his *Walks in South Lancashire* the radical Samuel Bamford presents ample evidence to suggest that much of the countryside around Manchester was still 'fair and green betwixt the towns' in the early 1840s.[49] However, after surveying the region from the nearby high moorlands he wrote, 'The smoke of the towns and manufactories is somewhat annoying certainly, and at times it detracts considerably from the ideality of the landscape'.[50] A decade later Leo Grindon, in his *Manchester Walks and Wild Flowers*, confirms the general thrust of Bamford's observations:

> Manchester itself, grim, flat, smoky Manchester, with its gigantic suburb ever on the roll further into the plain ... Manchester itself denies to no one of its three hundred thousand, who is blessed with health and strength, the amenities and genial influences of the country ... Yet is our town bosomed in beauty. Though the magnificent and the romantic be wanting, we have meadows trimmed with wild-flowers, the scent of the new-mown hay and the purple clover; we have many a sweet sylvan walk ...[51]

Although Grindon did acknowledge that the beauty of many of these 'sylvan walks' could only be accessed either by a long trek on foot, or by means of a railway journey.[52]

From the turn of the nineteenth century Manchester's smoke cloud was seen to be an ever expanding, ever present element of the urban environment. Dense black smoke billowing from the explosive growth of new factories and domestic hearths saw Leon Faucher compare Manchester to an active volcano in the 1840s, while Major-General Sir Charles Napier, appointed Commander of the troops of the Northern District in 1839, called Manchester 'the chimney of the world ... the entrance to hell realised'.[53] At mid-century Angus Bethune Reach, journalist for the *Morning Chronicle*, described his train journey to 'Cottonopolis' thus:

> The traveller by railway is made aware of his approach to the great northern seats of industry by the dull leaden-coloured sky, tainted by thousands of ever smoking chimneys, which broods over the distance ... Presently the tall chimneys begin to figure conspicuously in the landscape; the country loses its fresh rurality of appearance; grass looks brown and dry, and foliage stunted and smutty. The roads, and even the footpaths across the fields, are black with coal dust. Factories and mills raise their dingy masses everywhere around ... You shoot by town after

town – the outlying satellites of the great cotton metropolis. They all have similar features – they are all little Manchesters. Huge, shapeless, unsightly mills, with their countless rows of windows, their towering shafts, their jets of waste steam continually puffing in panting gushes from the brown grimy wall ... Some dozen or so of miles so characterised, the distance of course more or less according to the point at which you enter the Queen of cotton cities – and then amid smoke and noise, and the hum of never ceasing toil, you are borne over the roofs to the terminus platform. You stand in Manchester. There is a smoky brown sky over head – smoky brown streets all round long piles of warehouses, many of them with pillared and stately fronts – great grimy mills, the leviathans of ugly architecture, with their smoke-pouring shafts. [54]

As the colour of smoke depends on the type of coal being burnt, which can vary in hue from a dark brown to a deep black, Reach's depiction of Manchester's brown smoky air is, although somewhat unusual, likely to be accurate. A more conventional description of Manchester's dirty air is provided by Dr A. Emrys-Jones, a member of the Manchester and Salford Sanitary Association, who in April, 1882 wrote,

When travelling by rail, say from a romantic Derbyshire district to Manchester, we are forcibly struck by the wonderful contrast in the atmospheric condition. The beautiful, clear, blue sky is changed into a dark, murky, cloudy haze ... Everything that is calculated to please the eye is enveloped in a thick black veil.[55]

That the 'smoke nuisance' could be seen by contemporaries to be continually extending the boundaries of its territory may also be shown. In 1819, for example, Joseph Gregson told the Select Committee investigating steam engines and furnaces how twenty years earlier he had watched the ships sailing into Liverpool harbour from a hill in Toxteth Park. From the same vantage point, he apprised the committee, the mouth of the harbour was now obscured from view for most of the week, due to the increase in Liverpool's smoke.[56] In 1859 the scientist Robert Angus Smith recorded the extent to which Manchester's 'smoke nuisance' had advanced since the beginning of the century. 'The tinge of darkness in the atmosphere', he wrote, 'may be seen making a line of at least forty miles in length, and affecting the appearance of the sky and landscape'.[57] Rollo Russell, vice-president of the Royal Meteorological Society, stated that the dense volumes of smoke in the atmosphere over London meant, 'The sky, up to fifty miles distance at least, is more or less murky and ugly, the landscape blurred, and the horizon obscure.'[58] In July 1897 the Society of Chemical Industry held its annual meeting at Manchester, where the society's President, the chemical manufacturer Dr Edward Schunck, informed the gathered throng that smoke pollution in the city was visibly worsening:

I have no doubt that matters are much worse than they were years ago. In consequence of my dwelling occupying an elevated position [at Kersal], some 200 feet above the level of the River Irwell, in Manchester, I enjoy the advantage of surveying a considerable portion of the two towns of Manchester and Salford, and smoke and its emanation come in for a large amount of attention. That the amount of smoke has largely increased is evident from the fact that distant objects, such as the hills of Cheshire and Derbyshire, and even buildings on these hills, were formerly visible on clear days, and are now no longer seen on any occasion whatever.[59]

Moreover, Bamford's description of the minor annoyance smoke caused him when he looked out over the south Lancashire cotton towns in the 1840s can be contrasted with a botanist's view of the region's smoke from the high moorlands of Blackstone Edge in 1901:

> Deadly suburban fields form the most extensive element of the background; but what rivet the eye are the scores, and scores again, of mill chimneys, tall, straight, and lank, belching forth volumes of black, dense smoke straight at the rocks on which we stand! Rochdale, Littleborough, Bacup, Burnley, Nelson, Colne – each contributes its quota ... [and] the great smoke drift from South and East Lancashire [can] be seen crossing over the Pennine Range of moorlands and then mingling with the West Riding smoke.[60]

By the turn of the twentieth century the 'smoke nuisance' was not so much detracting from the 'ideality' of the landscape, as physically dominating it. Vistas of the industrial cityscape, and views of large swathes of the countryside surrounding Britain's northern factory towns, were monopolised by coal smoke. A little less than a century after Aston had made his confident assertion in print regarding the lasting quality of the rural views from Salford Crescent, T. Swindells was able to comment brusquely that Salford's early twentieth-century inhabitants would 'rub [their] eyes with wonderment' when reading it. Swindells then continued, 'As we see the pall of smoke that hangs over the valley to-day we realise the tremendous change that a century has produced.'[61] And while the smoke cloud expanded its sphere of influence farther and farther afield, the visual worlds of those who were trapped within the city's smoke seemed to contract. In June 1888 a resident of Ancoats, Manchester's main industrial district, wrote reprovingly: 'The atmosphere in this neighbourhood is so dense with smoke that it is impossible to see any object at a distance of a few hundred yards; and, as for sunshine, I have lived here ten years and never seen what could be called "brilliant sunshine."'[62] At times the 'smoke nuisance' was all that was visible to the naked eye in Manchester; and smoke made itself known to the senses in other ways too.

The conservationist and co-founder of the National Trust, Canon Hardwicke Drummond Rawnsley, penned the following lines about smoke pollution after a railway visit to Bolton and Manchester in 1890:

> ... chimneys, solid and square, were belching forth clouds of Erebean darkness and dirt, as if they had a dispensation from the Devil ... 'Moses Gate,' cried the porter, and we alighted. The heavens were black with smoke, and the smother of the mills, to one whose lungs were unaccustomed to breathing sulphurised air, made itself felt.[63]

On arrival, the visitor to one of northern Britain's industrial towns found that the acrid smoke cloud inhibited one's breathing and stung one's eyes, while the sticky 'blacks' soiled one's clothes and skin. The quality of Manchester's atmosphere was strongly criticised by a local doctor, who complained, 'The air we breathe, instead of being exhilarating and fresh, is a thick decoction of irritable and poisonous compounds.' [64] Moreover, in 1890 the socialist Edward Carpenter described atmospheric conditions in Sheffield thus:

> A foreigner, walking with me one day through the streets of Sheffield, said, 'Well, I never was in a place before where the dirt *jumped up and hit you in the face!*' ... the common 'black', with which we are all so pleasantly acquainted, which hits us, as my friend said, so playfully in the face, or descends so gracefully upon the tip of our nose; ... sometimes showers down like rain in the streets of our great cities, or at other times, mixed with actual rain, paints our clothes, our faces, ... in that funereal colour which is supposed by frivolous foreigners to accord with our natural temperament ... [65]

Carpenter's sarcastic tone reveals the great annoyance he and other contemporaries experienced as the 'smoke nuisance' more than lived up to its name. That the irritation caused by 'blacks' and smoke was indeed acutely felt can be illustrated by the remarks of the president of the Institute of Sanitary Engineers, Arthur J. Martin, who exclaimed, 'We all swear like troopers ... when it gets into our eyes'.[66]

Smoke could also be perceived outside of the city as more than simply a darkening of the sky overhead. The engineer W.C. Popplewell documented the unpleasant sensation he experienced when he touched the oily, blackened vegetation that surrounded the industrial towns.[67] In addition, Edward Carpenter commented indignantly, 'I have *tasted* the sulphury exhalations of Sheffield twenty miles off, in the heart of the Derbyshire Peak.' [68] In 1902, addressing a sanitary congress held at Manchester, the physician and psychologist Sir James Crichton Browne interwove many of these unpleasant sensory perceptions of the 'smoke nuisance' when he declared,

> This is the age of smoke in which we are living. ... You are smothered in the products of combustion all the year round; in winter these settle down on you

as fogs, grim and horrible, and you must go far afield before you can look on a sheep with a white fleece or pick a flower or blade without soiling your fingers. A sable incubus embarrasses your breathing, a hideous scum settles on your skin and clothes, a swart awning offends your vision, [and] a sullen cloud oppresses your spirits ...[69]

With regard to odour, which Browne omitted, German contemporaries even spoke of a characteristic *Englischer Geruch* – a smell of coal.[70] The 'smoke fiend', an alias by which coal smoke was often known, was a constant – and highly disagreeable – presence in the everyday lives of all city dwellers: a presence that nature could not be relied upon to disperse.

The influence of climate and topography

Until very recently it was thought that the immeasurable bounds of the earth's atmosphere was an inexhaustible sink that could easily absorb, dilute, and ultimately neutralise all the pollutants that industrial chimneys could emit. The tall chimneys, some well over 300 feet high, that dominated the townscapes of the industrial north were designed in part to reduce local air pollution by discharging smoke as far up into the atmosphere as possible, where it would then be transported well away from its point of origin by strong air currents.[71] The awe with which contemporaries regarded the perceived immensity of the earth's atmosphere is captured in the words of the meteorologist Rollo Russell in 1896:

> The atmosphere has been compared to a great ocean, at the bottom of which we live. But the comparison gives no idea of the magnitude of this ocean, without definite bounds, and varying incessantly in density and other important qualities from depth to height and from place to place. Uninterrupted by emergent continents and islands, the atmosphere freely spreads high above all mountains and flows ever in mighty currents at levels beyond the most elevated regions of the solid earth.[72]

Nature was erroneously thought to have unlimited powers to purify itself of all noxious substances that polluted the air: a notion that did little to discourage the extravagant and wasteful use of coal. For example, in 1901 the Nobel prize winner Clemens Winkler had insisted, 'The volumes of consumed coal disappear without a trace in the vast sea of air.' [73] That the 'mighty currents' of the atmosphere were thought to be more than capable of effectively nullifying localised air pollution around Manchester can also be demonstrated. In a paper delivered to the Manchester Literary and Philosophical Society the alum manufacturer Peter Spence argued for the construction of one huge chimney that would replace all of city's many thousands of industrial and domestic chimneys. It was to stand 600 feet in height, being 100 feet in diameter at its top and 140 feet at its base, with brickwork some 10 feet thick. This immense chimney, Spence believed, would successfully remove from the city the combustion products of the two million tons of coal it burned annually by,

> ... separat[ing] them entirely from that portion of the atmosphere in which we live, and ... convey[ing] them into the vast atmospheric ocean over our heads at a point which would effectually preclude their returning upon us.[74]

However, his ambitious plan to rid Manchester of smoke, (which also utilised the city's sewerage system), was never acted upon; although the construction of huge twentieth-century power station chimneys hundreds of metres high was unquestionably based upon very similar principles. But, as Spence's words intimate, coal smoke did regularly descend from Manchester's numerous smaller chimneys to taint the air in which its inhabitants lived. Constant atmospheric and climatic changes meant that the smoke did not always rise up into the abundant 'ocean of air' and simply disappear harmlessly as many contemporaries wished. In 1854 Robert Angus Smith described some of the ups and downs associated with the universally adopted method of utilising tall chimneys to remove smoke:

> When the sky is open, there is a fine clean air in our streets; the gases seem rapidly to follow the laws of their diffusion, and they leave at a rapidity which almost satisfies theory. There is then, no doubt, a constant flow of air into the town, along all the streets and roads, to make up for that great current which is continually rising into the atmosphere. If the air were constantly clear, we should then have very diminished evils of the kind, as this constant ventilation of the town would take place; but when it is clouded and moist, an entirely different state of things occurs. The acid and other impurities become dissolved in the moisture, and the black parts of the smoke become wet and heavy. At this time the air becomes very acid, and the atmosphere, as we approach the more crowded parts of the town, becomes sensibly deteriorated. This must have been observed by many. If the day be foggy, this takes place in still greater force, and these floating particles of liquid must have a strong influence on those who are subject to coughs, or are otherwise delicate, being as they are, solutions of acid and acid salts. This is one of the ways in which we suffer from a moist climate.[75]

Smith had observed that during favourable weather conditions working industrial furnaces and household fires created by their actions a strong, bracing circulation of the air in Britain's largely unplanned, densely packed factory towns. This was believed to positively benefit the health of Britain's townspeople by permitting them to breathe cleaner air being drawn in from the surrounding countryside. For as the warm smoke drifted away from the town it pulled in currents of cooler fresh air behind it, thereby revitalising the vitiated urban atmosphere. The health of London's populace was thought to have improved due to this process. In 1819 a witness informed the Select Committee on steam engines and furnaces, 'I am aware it is allowed the health of this metropolis has increased since the use of coal; a brisk coal fire, causing a great circulation of air, promotes health.' [76]

However, Britain as a whole, and Lancashire in particular, is not generally blessed with long periods of clear and dry weather at any season of the year. Although there are considerable differences in Britain's annual rainfall from year to year, the western side of the island usually receives over 40 inches per annum, with rain falling in these areas on more than 200 days of the year.[77] Indeed, the humid climate and cloudy skies of Lancashire are often stressed as important natural advantages when explanations for the growth of the cotton industry in the county are sought.[78] A moist and overcast atmosphere was especially favourable for working with this exotic fibre. But, as Smith notes above, the same humid climatic conditions were also a very serious disadvantage when it came to the dispersal of the region's intensifying smoke cloud. Moreover, there were other vicissitudes of the weather and the physical environment that intertwined with the 'smoke nuisance' to affect the lives of Manchester's inhabitants and the growth of the city itself.

The climate of the region must be considered in relation to local topography. Most local urban air pollution problems were caused by high concentrations of smoke at low atmospheric levels, often no more than a few hundred metres above the ground. It follows, therefore, that local relief – in the sense of variations in the elevation of the earth's surface – will be an important contributory factor regarding the distribution of air pollution. As many of Lancashire's cotton towns were situated within the county's upland river valleys, when periods of calm weather prevailed, with little or no wind to carry off pollutants, the tall factory chimneys often failed to lift the smoke above the tops of the surrounding hills. The smoke cloud did not always find its way unhindered across the Pennine Range from Lancashire to the West Riding of Yorkshire, or *vice versa*. Particularly during the winter months, the sluggish air conditions associated with anticyclonic calms saw inversions of atmospheric temperature occur that sometimes trapped a town's smoke in the streets for several days at a time. The reversal of normal temperature variation in the troposphere, where temperature usually decreases with increasing altitude, confined high concentrations of pollutants to the layer of air immediately above the surface of the town or city. In industrial districts where large amounts of coal smoke were emitted into the atmosphere, black and brownish 'lakes' of accumulated air pollution were formed that lingered to smother their inhabitants. Radiation inversions were also quite common in urban areas on winter nights, as the ground cools more rapidly than the air above. When inversions develop they inhibit the dispersion of smoke vertically as a layer of warm air, sometimes no more than a few tens of metres above the ground, acts as a horizontal lid, effectively bottling up a town's atmospheric impurities within the valley's sides. The colder, heavier air beneath this warm 'ceiling' has no inclination to rise until more turbulent air flows return or the sun breaks up the inversion by warming the valley floor.[79]

The site of Manchester itself is flat and characterless, with some parts of the city being little more than 30 metres above sea level. Located within the Mersey river basin, Manchester is, however, bounded to the north, north west, east, and south east by the Pennines and their foothills. The girdle of hills to the north and east rise to an altitude of over 300 metres above sea level, although those to the south are much lower.[80] Given the appropriate weather conditions, the smoke of Manchester could and did remain firmly entrenched within the encircling uplands for long periods. Moreover, the monotonously flat Manchester lowland presented few natural obstacles to check the constant drift of smoke into the city from its many surrounding satellite towns – especially Salford. Under such circumstances it is difficult not to have some sympathy with Shena Simon's lament that 'Nature has not been kind to Manchester in helping her to get rid of smoke.' [81]

Fogs also formed readily in the smoky towns and cities as the particulate pollution – smoke and soot – provided in abundance the necessary nuclei for condensation and the formation of water droplets, especially at low temperatures. In nineteenth-century Manchester fogs grew denser and gradually became more frequent, particularly during the winter months, as a result of the presence of the 'smoke nuisance'. In cold, calm atmospheric conditions fog often went hand in hand with an inversion, with the smoke also retarding the dissipation of the dense choking fog in urban areas. The dark, tarry pollutant matter formed a sticky film around the water droplets, which meant that they evaporated far less easily in the rays of the sun. In addition, it was the oily, sulphurous impurities that gave urban 'pea-soupers' their characteristic taste and smell, as Robert Angus Smith detailed:

> ... a still and otherwise peculiar state of the air, such as that which brings fog, causes the accumulation [of smoke] to a very wonderful extent, and increases the intensity of that phenomenon. It is then that we perceive how acrid the substances in smoke may show themselves. We may then smell sulphur, that is sulphurous acid, distinctly; we may obtain from the air comparatively large amounts of sulphuric acid; and we may see minute globules of liquid which are really dilute vitriol. These affect the eyes and throat, even before the smell. [82]

George Davis, District Inspector of Alkali Works, considered Manchester's fogs to have a particularly penetrating smell when contrasted with other towns and cities. 'The peculiarity of the Manchester fog', he wrote, 'as compared with the fogs of other places – with London for instance – is its extreme pungency, its unusually high charge of sulphurous acid.' [83] Whenever the wind died away and the temperature dropped, the suffocating fog, exacerbated by the 'smoke nuisance', remained trapped at ground level and caused considerable discomfort to city dwellers in many other ways besides irritated eyes and burning throats. As day was turned into brumous night, traffic in the busy city streets ground to a halt as visibility dimmed sharply, trains were delayed, the number of accidents and crimes were believed to increase, and trade suffered as 'the affairs of a great city are disarranged'.[84] In January 1888, absolute darkness had threatened to hold sway in Manchester for a time, as a long-lasting period of fog tested the city's gas-producing powers to their very limit. At Christmas 1904, during four days of heavy black fog, Manchester's gas supply was exhausted, leaving people to blindly make their way through the city streets as best they could.[85] But although fogs were expensive in terms of lost time, lost revenue, and an increased use of fuel for heating and lighting purposes (which in turn aggravated the problem), it was argued that they did have some redeeming qualities. In December 1905, the *Evening Standard & St. James Gazette* proclaimed that London without its 'particular' would be 'unthinkable', adding:

> One of the ... great advantages of fog is that it varies the public talk. Another is its indulgence of the feeling for adventure, felt more or less by every true-born Briton. A journey into a fog is an adventure into the unknown. It is the only chance of adventure many of us ever get.[86]

However, while a small minority of contemporaries may have enjoyed venturing out into dense fogs as a 'natural' source of excitement and adventure in the city, or as a stimulus to aid conversation, there is little evidence that attests to their great popularity amongst the urban masses.[87] There is, in fact, a great quantity of contrary evidence (discussed below) to show that many Britons were anxiously engaged in discussing its more harmful effects, such as reducing sunlight in urban areas or the damage it caused to the health of city dwellers.

The incidence of fog in Manchester, though it varies from year to year, is generally most frequent through the autumn months and the early part of the winter, peaking in December or January and thereafter declining.[88] In addition, although figures are thin on the ground, the frequency of Manchester's 'stinking fogs' seems to have increased during the nineteenth and early twentieth centuries. There had been 24 fogs recorded between the years 1804–10; 52 between the years 1820–25; and by 1900–01 the number of dense fogs in Manchester had risen to match this number, again reaching 52 but in the space of just two years.[89] In Manchester fogs were usually 'most suffocating and noxious' close to the banks of the River Irwell between Cornbrook and Blackfriars, followed closely by areas next to the River Medlock.[90] And as 'ordinary' thick white fog forms naturally over rivers, streams, and expanses of open water, contemporaries often pointed out that towns and cities like

Manchester would never be entirely free from fogs. But the increasing frequency of Manchester's fogs, their unnatural brown or black colours compared with white 'country' fogs, their sulphurous smell, and the acidic nature of these dense choking visitations, were widely recognised to be the result of burning enormous quantities of bituminous coal. The formation of smoke-saturated urban fogs was viewed as being anything but a natural occurrence, as the *Manchester City News* noted in 1884: 'Fogs, such as we have had in Manchester, are products of civilization which admit of mitigation and abatement.' [91] During the early part of the twentieth century fogs were observed by contemporaries to be declining significantly in London.[92] However, in Manchester the incidence of foggy days actually rose by 30 per cent and the number of 'gloomy' days by 100 per cent during the same period.[93] The calm, cold weather conditions which produced urban fogs, especially prevalent during Britain's winter months, were by the early years of the twentieth century recognised to be potentially catastrophic for every industrial city in the land.[94] With high levels of air pollution existing in urban areas the natural cleansing of the atmosphere by revivifying winds could not afford to become stalled for long by calm weather, as James Russell, Medical Officer of Health for Glasgow, emphasised:

> Whenever the scavenging of the air is interrupted by calms, so that the smoke products accumulate, the atmosphere of our streets thickens and … at mid-day we have mid-night. Smoke not only loads fog with impurities but tends to produce fog. When, therefore, in the winter we happen upon a low temperature, a high barometer and a dead calm, we have an arctic night with a mephitic, irresponsible atmosphere in which we move about choking, our eyes irritated, our faces grimy – 'a purblind race of miserable men.' It is then that our pulmonary mortality is run up, and the death-rate of the year determined.[95]

Contemporaries, then, were becoming aware of the dangers posed by certain configurations of the weather when wedded to the inefficient combustion of fossil fuel. Indeed, in 1902 W.N. Shaw, a member of the Sanitary Institute of Great Britain, declared, 'On foggy days when we thrust our refuse into the atmosphere, it simply descends upon our heads and into our houses. We might as well have no chimneys at all.' [96] But climatic conditions were constantly changing, and for much of the year the currents of the 'atmospheric ocean' were strong and turbulent enough to displace the bulk of Manchester's smoke at least as far as the moorland hills and neighbouring towns to the north and east of the city.

If Victorian Britain's urban dwellers depended upon breezy air conditions to keep the smoke moving in their polluted towns, the direction of the prevailing winds was equally important to the way in which an industrial city developed physically. In Manchester the winds blow from all points of the compass, but with the prevailing and strongest winds blowing from the south west. This meant that when the dense sulphurous smoke left Manchester's tall chimneys it usually moved north east, and this was to have a marked effect on the shaping of the city. When the directors of Manchester's Botanical Society were searching for a site for their gardens they wished to acquire a spot that was free from smoke, so they approached the eminent scientist John Dalton for help as he had long been studying the region's winds and weather. In 1827 the Botanic Gardens were established on a sixteen acre site at Old Trafford, to the south of Manchester, after Dalton's calculations had shown that the wind would blow from the city towards the gardens for only about five weeks of the year.[97] As the smoke nuisance gradually worsened the more prosperous middle classes left Manchester to take up residence in suburbs that were also situated well to the south of the city, such as Alderley Edge, Altrincham, and Wilmslow. With time Manchester's factories and industries gravitated towards its north-eastern boundaries, in part because of the high cost of land in the increasingly commercially orientated city centre, but also because it was known that the smoke tended to disperse from the city more efficiently at this point.[98] But Salford's industries likewise sited themselves on the north-east boundary of the city along the banks of the Irwell, and not simply for the close proximity of a useable water supply. As Manchester's combustion products moved away, Salford's smoke rolled in to replace it – a source of much antagonism between the twin cities. The poorest city dwellers were forced to live amongst the mills and factories in north-easterly districts like Ancoats, while the better-paid among Manchester's working classes might at least escape the worst of the smoke by residing in the more southerly situated Chorlton-on-Medlock or Hulme.

When giving his evidence to the Select Committee of 1843, the Reverend John Molesworth presented its members with a map of Manchester that specifically emphasised the importance of smoke and the prevailing winds on the city's residential patterns:

> The object of it is to show in what parts of the town of Manchester the smoke principally prevails, from the prevailing winds; and I believe the conclusion [is] ... that all the wealthy manufacturers themselves resort to one particular part of town, which is a part in which the smoke does not generally prevail; which is an acknowledgement, on their parts, of the greatness of the nuisance ... Many more reasons may be assigned for the nonresidence of the large manufacturers in this district. But it appears to me very remarkable, that while so many of the tradesmen live out of town (for hardly any live in Manchester) in other directions than those in which the vast mass of smoke is carried, few reside in the outskirts on the eastern part of the town.[99]

Molesworth's map showed that most of Manchester's affluent businessmen had made their homes beyond the reach of the worst of the smoke, while the city's poorest inhabitants were housed directly in the smoke cloud's path. Moreover, unlike the strong, prevailing south-westerly winds, the less frequent gusts from the north and east are usually fairly weak, which makes them less favourable for the dispersion of air pollutants.[100] The atmosphere is an exceptionally dynamic system that is influenced by changing meteorological conditions, both on a daily and seasonal basis. Manchester's winds, for example, are usually stronger during the afternoons than at night, with extreme gusts being more frequent in winter than in summer.[101] Air conditions in Manchester were variable; therefore the 'smoke nuisance' did not remain at a constant level in the atmosphere. Nonetheless, city planners could be certain that north-easterly districts such as working-class Ancoats would bear the brunt of the smoke for much of the year. However, along with the gusting winds, rain and snow were the other major natural scavengers that could greatly reduce urban air pollution – at least for short periods of time.

If in large towns the method of leaving the purification of the air to the natural rhythms of atmospheric currents failed from time to time, heavy showers of rain or snow often helped to reduce the density of the 'smoke nuisance' in urban areas. Driving rains or prodigious snowfalls temporarily cleansed the atmosphere of smoke by removing the malodorous pollutants as they fell, as Manchester's Edward Schunck described:

> Of course, the contamination of our atmosphere by coal smoke is not at all times equally offensive. In dull, close – particularly foggy – weather, whether in summer or winter, it is peculiarly objectionable, the more so since at such times the carbon of the smoke is mixed with sulphurous acid from the pyrites of the coal. On the other hand, during or after rainy weather, the atmosphere in Manchester, and perhaps in other places, is comparatively pure and pleasant, the impurities being washed down in the rain.[102]

But, as Edward Carpenter outlined above, getting caught outdoors during such a downpour could be a very unpleasant experience. Contemporaries reported numerous instances of 'black rain' and 'black snow', often at considerable distances from large industrial towns. In 1881 a fall of black snow (*smudsig snefeld*) in Norway was attributed to industrial emissions originating from the manufacturing districts of Great Britain.[103] Closer to home, by the turn of the century rain charged with soot was frequently observed to fall in the Lake District, leaving the shores of its lakes fringed with a greasy black scum.[104] Indeed, the work of the scientist Richard Battarbee has clearly shown that many of Britain's lakes started to acidify in the mid-nineteenth century due to the increase in air pollution.[105]

For much of the nineteenth century, contemporary records and accounts of Britain's changing weather patterns are as fragmentary as those for urban fuel consumption. But some scientific research and data concerning the climate and weather of nineteenth century Britain and Manchester in relation to smoke does exist to put into context the personal experiences of air pollution outlined above. Wilfrid Irwin, a member of the Society of Chemical Industry, after analysing a heavy fall of snow in Manchester during February 1902, calculated that more than three hundred tons of soot had fallen on the city over a ten day period, or over thirty tons per day.[106] Flurries of snow are actually more efficient scavengers than showers of rain due to the high surface area of the flakes. However, the results of an investigation comparing the impurities found in rainwater collected in several British towns and cities during October 1914 were said to show that air pollution was less severe at Manchester than in Oldham, Birmingham, London, Sheffield, and Bolton. Oldham was declared the dirtiest town of those tested, recording 29 metric tons of soot, tar, and dust deposited per square kilometre for the month, against 23 for Birmingham, 22 at London and Sheffield, 21 for Bolton, with just 20 metric tons per square kilometre at Whitworth Street in Manchester.[107] However, figures for soot, tar, and dust deposits recorded by the Manchester Air Pollution Advisory Board for January 1915 puts this ranking order of insalubrity in some doubt, as Table 5 shows. The striking difference in the total amounts of impurities recorded at the collecting stations in Fallowfield and Ancoats Hospital clearly demonstrates how levels of air pollution can vary significantly from place to place within a single city like Manchester. Where such stations should be sited in Britain's cities is still a matter that is being hotly debated today. But if the figures utilised had been for industrial Ancoats rather than the more commercial, centrally-located Whitworth Street, Manchester might well have run Oldham very close for the title of Britain's dirtiest town in 1914.

Collecting station	Metric tons per sq. km.
Ancoats Hospital	30.59
Philips Park	22.59
Whitworth Street	22.51
Queen's Park	20.18
Moss Side	18.69
Whitefield	15.53
Fallowfield	13.24
Davyhulme	12.68
Cheadle	10.63
Bowdon	6.25

Table 5. Soot, Tar, and Dust Deposits for the Manchester Region in January 1915.

Source: *First Annual Report of the Sanitary Committee on the Work of the Air Pollution Advisory Board,* City of Manchester, 1915, p.18.

Moreover, the Air Pollution Advisory Board's figures for sulphate deposited in the city between August and December 1914 are very high when compared to today's values for urban sulphate concentrations. The results of the Board's scientific inquiries varied from a low of 39 grams of sulphate per square metre in August to a high of 171 grams per square metre in December. The usual value range today is from 1 to 6 grams of sulphate per square metre per month in Britain's cities.[108] Sulphur, present in quite high concentrations in Lancashire coal, is emitted into the atmosphere as sulphur dioxide during the process of combustion and is ultimately oxidised to sulphuric acid and sulphate salts. Although sulphate enters the atmosphere naturally from volcanic eruptions, the biological decay of organic matter, and wind-blown sea-spray, such high urban concentrations can only have had their origins in the region's large-scale emissions of industrial and domestic coal smoke. And this 'excess sulphate' contributed in no small degree to Manchester's burgeoning acid rain problem.[109]

From the late 1840s onwards Robert Angus Smith's pioneering work on the air and rain of Manchester identified coal combustion as the principle cause of the great acidity of the city's rainfall. For example, in 1876 he noted,

> In Manchester, we have rain containing nearly a grain of free sulphuric acid per gallon. Where my laboratory is, by no means in the centre, the rain reddens litmus as it falls as rapidly as vinegar does, and trees and shrubs refuse to grow, even grass looks unhappy ... We do not require chemical works to destroy trees; coal alone is sufficient, although slower, whenever chimneys are sufficient in number to produce the acidity spoken of ... and all the coal districts shew this abundantly.[110]

By 1872 Smith had articulated many of the concepts that are now part of our present understanding of acid rain. Using Smith's extensive chemical analyses of free acids in Manchester's rainwater, Dietrich Schwela has recently converted his figures to estimate that the pH value of the city's rainfall over a century ago was a very low 3.5. This is a far more acidic value than today's measurements of around pH 4–pH 4.4 for acid rain in Europe and the USA.[111] The available scientific data, then, far from contradicting the lay evidence of contemporaries like Canon Rawnsley and Edward Carpenter, essentially lends considerable support to their vivid descriptions of the highly polluted atmosphere of Britain's industrial cities.

By the late nineteenth century the wisdom of leaving the efficient dispersal of the urban smoke cloud to a system which relied on harnessing the forces of nature was beginning to be seriously questioned. For example, W.N. Shaw believed that cleansing the urban atmosphere of smoke was an undertak-

ing that was increasingly beyond the powers of nature to perform successfully
– even on windy days:

> Not only is the magnitude of the task too great but the manner in which the
> atmosphere deals with it is not by any means satisfactory. It does not consume
> or annihilate the smoke, or render it harmless as it does the germs of many
> diseases; it does not even, like well-behaved scavengers, carry it to a suitable spot
> and deposit it in neat or untidy heaps; the air carries its load a little way, longer
> or shorter according to the state of the weather, and then drops it regardless of
> consequences.[112]

The inexorable growth of the smoke cloud saw numerous schemes hatched in
the second half of the century that were designed to augment the waning powers
of nature to neutralise the 'smoke nuisance'. That some contemporaries felt that
nature was losing the battle against smoke pollution may be exemplified by
briefly sketching a selection of the most inspired and ingenious plans. Not
forgetting Spence's colossal chimney, these plans included: constructing large
electrically driven fans that would blow away the smoke from a city's chimneys;
delivering massive electrical discharges into the atmosphere to stimulate artificial rainstorms; and, not least of all, installing an extensive system of aerial water
sprinklers within a town to wash out the suspended particles of carbon.[113]
Industrialisation and urbanisation, then, had not so much mastered nature as
gradually and palpably overburdened it locally with air pollutants. In 1876
Robert Angus Smith protested that the atmosphere had become 'a great source
of misery' to the urban masses. 'The air outside which we know as a reservoir of
purity and an agent of purification', he wrote, 'becomes to them a thing to be
dreaded, being full of soot and coal dust.' [114] So, what were the consequences of
overloading the atmosphere with coal smoke?

Blackening the face of nature

Over half a century before the onset of the Industrial Revolution, Daniel Defoe
described the approach to the 'greatest meer village in England' in the following
uncomplimentary terms:

> ... on the road to Manchester, we pass'd the great bog or waste call'd Chatmos
> ... The surface, at a distance, looks black and dirty, and is indeed frightful to think
> of, for it will bear neither horse or man ... What nature meant by such a useless
> production, 'tis hard to imagine; but the land is entirely waste ... [115]

Like Defoe, many Victorians were highly critical of Manchester's natural
environmental attributes, primarily because of the flatness of the city's
uncommanding position on a great plain, allied to an absence of picturesque

waterways. If nature had not been kind to Manchester in helping to disperse its smoke, the aesthetic value of the countryside in the immediate proximity of the city was also thought to be very limited as a result of nature's 'ungenerosity'. In 1859 the distinguished Manchester botanist Leo Grindon wrote plaintively of the city's lack of impressive surroundings:

> ... we have no grand scenery: no Clyde, no Ben Lomond, no Leigh Woods, no St. Vincent's Rocks, no Clevedon, no Durdham Down; our rivers are but streamlets, oftentimes anything but limpid; our hills and mountains belong to our neighbours rather than to ourselves; our lakes, with one or two exceptions, the beautiful waters of Tabley and Rostherne, are but reservoirs, indebted less to nature than to art; our waterfalls are only in our portfolios.[116]

Manchester was thought to compare very unfavourably with a host of Britain's busiest cities in this respect. The 'dignity and charm' of the site of nearby Liverpool, for example, with its broad estuary and port, was grudgingly acknowledged to be aesthetically superior.[117] In the absence of awe-inspiring views, Manchester's vegetation became doubly important to those such as Grindon who took great pleasure in the sublimity of the natural world:

> ... however wanting in attractions to the mere adulator of 'fine scenery', every little flower, every bend of branches, and sweet concurrent play of light and shade, every pendent shadow in the stream, becomes animated with a meaning and a power of satisfying such as none but those who accustom themselves to look for it *here*, can find in the most favoured and spacious landscape ... nothing less than total nakedness of surface can take from a place its power to interest and please.[118]

By degrees, however, air pollution deprived the city's flora of its 'power to please', slowly obliterating the vestiges of natural beauty that Manchester was thought to possess. The 'smoke nuisance' blackened the face of nature in the city, and by the last decades of the century Manchester was virtually barren. In 1898 the Very Rev. Dr L.C. Casartelli, President of the Manchester Statistical Society, protested against nature's evanescence: 'if Nature has been thus chary of her charms, or rather denied them to us almost entirely, man has been busy to spoil and destroy even what little has been allowed us'.[119]

Unregulated urban sprawl, and an enormous increase in population, badly affected Manchester's plant life by putting cultivated, open spaces at a premium. In common with many other large towns and cities of the period, green spaces and private gardens for recreation were wanting. In 1844 Friedrich Engels noted how the builders in Manchester had 'built up every spot' to provide housing for the new urban masses.[120] Air pollution, not overcrowding, was responsible for destroying the modest amount of vegetation that remained, as the *Manchester Guardian* explained in 1842:

We hear and read much of the smokiness of the metropolis; and the complaints made on that score are certainly far from being groundless: but, if there is good ground of complaint against London, what must be said with respect to Manchester? Not only is the greater density of smoke in this town palpable to the senses, but it is most unequivocally manifested by the different effects produced on vegetation in the two places. Every man familiar with London is well aware, that there are many trees not only growing, but flourishing, in some of the closest and most confined quarters of the city of London; and that the squares at the west end of town are planted with trees and flowering shrubs in very fair condition: whilst, in one of the openest spaces in the interior of Manchester, – namely, the garden of the infirmary, – flowering shrubs will not grow at all; and of a long line of trees which were living some years ago, the one or two which alone are left are in the last stage of decline and decrepitude; – clearly showing, that in this town the air is vitiated to a far greater extent than in the metropolis.[121]

The smoke-laden atmosphere banished vegetation from the heart of Manchester. For anyone who lived in the centre of Manchester in 1800 fields and meadows were only a short walk away. Even at mid-century there was still no lack of green pleasant places on the outskirts of the city. But the rising tide of buildings and people continually encroached upon the countryside adjoining the city, which meant that its poorer inhabitants found their opportunities for spending a day out in the open air were gradually eroded away year by year. By the end of the century Herbert Philips, Chairman of the city's Committee for Securing Open Spaces for Recreation, was moved to highlight the contrasting experiences of two sets of boys, the 'working lads' of Manchester and those of Macclesfield:

To the former such a thing as [a] country ramble is the rarest experience. For miles the wilderness of streets spreads round them, and a skirmish across the fields and over the hillsides on Saturdays – a common incident in the life of a Macclesfield lad – is for Manchester boys an absolute impossibility, a thing undreamt of.[122]

Fresh air, green fields and a clear sky had largely become 'meaningless terms' for the urban masses.[123] The divide between town and country – a false antithesis for many environmental historians – is valid in at least one respect: large numbers of city dwellers rarely visited rural areas or experienced nature intimately.[124] However, since the 1840s concerted efforts had been made to provide green open spaces in Manchester to compensate for the loss of urban vegetation and easy access to the surrounding countryside. Public parks and playgrounds were established, and many other available spaces, such as church-yards and cemeteries, were planted with trees, shrubs, and flowers to provide the city's weary workers with relaxing and revivifying areas of greenery.[125] In 1846 the first public parks were opened in Manchester and Salford: Philips Park at

Bradford, Queen's Park at Harpurhey, and Peel Park at Salford. A decade later *Bradshaw's Guide to Manchester* stressed the important contribution these new green oases had made to urban life. 'They are the weekly resort of tens of thousands of our hard-working labourers and artizans', its author noted, 'and form that most agreeable variety of something like green fields and shady trees in the vicinity of Manchester smoke.' [126] Manchester's parks were envisioned as the health-giving 'lungs' of the city, places where its jaded citizens could go to unwind, take exercise, and breathe purer air. In addition, it was thought that they could also follow the passing of the seasons, and perhaps catch an unspoilt view of the sun and sky. The reality was that Manchester's 'breathing spaces' suffered from the same pollution as did the rest of the city. Factory chimneys often ringed public parks, and in 1870 a special sub-committee was appointed by Manchester Corporation to investigate the influence of air pollution on the vegetation of Philips Park and that of the nearby cemetery. It reported,

> So considerable was the quantity of smoke sent through the Park on one of the visits from these works that the atmosphere was completely clouded by it, and the smell of the smoke was stifling. It is quite impossible that healthy vegetation can subsist in such atmospheric conditions, and the trees in the higher portion of the Park were severely suffering. [127]

A little over a quarter of a century later Herbert Philips announced that their struggle for life was over: 'in the older parks forest trees have ceased to exist. The sulphurous acids from our chimneys have destroyed them'. [128] In the end it was only through constant human intervention that a modicum of vegetation survived at all in smoky Manchester.

From the mid 1850s onwards, in order to help combat the damage caused by the 'smoke nuisance', contemporaries had started to pay very careful attention to the type of trees and flowers they chose for cultivation. [129] Beauty, however, was not the prime consideration in the selection process, but a proven ability to withstand the severely polluted conditions of the city. In 1886 Mrs Haweis produced a notable book entitled *Rus in Urbe: or Flowers that Thrive in London Gardens and Smoky Towns*, in which she catalogued in meticulous detail those plants which would, and those which would not, grow in a smoky atmosphere. 'High-class roses', she wrote, 'will not thrive, and it is of no use to try them, even under glass; the velvet leaves cannot throw off the soot in the air, the pores are choked, and death is inevitable.' [130] Keith Thomas has estimated that as early as 1826 the rose, 'the queen of flowers', was available in no fewer than 1,393 different varieties in Britain. [131] But, by the last quarter of the century many kinds could no longer be successfully cultivated in industrial towns. For example, in 1882 Leo Grindon wrote despondently that the roses in his garden had 'all slowly departed'.[132] Based on long-term personal observation, Grindon also produced a list of 'friends to fall back on in a time of smoke', while advising his readers that,

> ... in matters of this kind we must learn to content ourselves with what *will* grow and live, not what we think would look nice and ornamental. If we cannot have that we like, the first rule of prudence is to cultivate a disposition to like that which we have.[133]

Vegetation in the Manchester area began to be dominated by flora proven, by trial and error, to be capable of enduring the smoke: and among the most common species to be found in the city were poplar trees, privet, elder, ash, hawthorn, honeysuckle, lupins, rhododendrons, and willows. By the 1880s the notion of cultivating plants that were 'appropriate' for the polluted environmental conditions was well established, as the following contribution to the *Manchester Guardian* exemplifies:

> ... even under existing circumstances it is quite possible to increase our 'verdure' by planting in every available space the large-leaved, or balsam poplar. 'Tis true they will not live many years in dense smoke, but they are the best kind for resisting it, and will hold out for different periods up to about ten years, and by eliminating the dead ones and putting fresh plants ... in their place, patches, clumps, and avenues of greenery might be produced and sustained throughout the summer to the delight of thousands who would look upon them ... the best evergreen to plant would be the *aucuba japonica*, as every shower would wash the soot from its glazed and spotted leaves, and its foliage is compact and of a good colour.[134]

Indeed, that there were fine shows of flowers to please the eye of visitors to the city's public parks depended almost entirely upon the skills and perseverance of head gardeners rather than on natural growth. Dead and dying vegetation had to be constantly exchanged for young healthy plants raised in outlying nurseries. Rhododendrons, for example, survived for only three years in Philips Park, and by their third year they were 'anything but beautiful to look upon'.[135] By the early years of the twentieth century the Manchester Parks Department had established a nursery on 65 acres of land nine miles outside the city in Cheshire. It was kept stocked with 'hundreds of thousands' of trees, shrubs, and bedding plants, in various stages of growth, ready for planting out in parks and playgrounds.[136] In 1907 alone the nursery supplied 26,800 trees and shrubs and around 70,000 herbaceous plants for use in Manchester's parkland. In the first week of July each year, in an effort to restore greenery to the city centre, about 1,000 trees and shrubs in tubs were also placed in front of Manchester's public buildings and in its squares.[137] But these 'tub forests', like a flock of exotic migratory birds, stayed to brighten the city streets only during the summer months. By the end of September their foliage was so damaged and discoloured that they had to be returned to the country to recuperate.

Illustration 2. Trees Damaged by Air Pollution at Salford, July 1857.

Source: Salford Local History Library.

The concept that public parks could mimic the natural rhythms of the seasons for city dwellers, or provide uplifting, healthy vegetation seems, given Manchester's atmospheric conditions, to have been one of somewhat limited worth. Few plants pushed their green shoots through the blackened earth spontaneously in the springtime: they were transported into the city in bulk from the countryside. The growth and vitality of these specially selected 'resistant' species was then retarded by heavy sootfalls, acid rain and reduced light intensity. In summer, for example, the colours of city flowers were observed to be paler than flowers of the same type blooming in outlying areas.[138] At Leeds the extensive studies of scientists Julius Cohen and Arthur Ruston, published in 1912, found that the vegetation cycle was greatly disturbed by smoke. In the autumn ash trees in the purer parts of Leeds retained their leaves for up to six weeks longer than ash trees in the more contaminated districts of the city.[139] The variegated foliage of mature forest provided one of the principal autumnal attractions for contemporary nature lovers. But a thick black deposit of tarry soot on the leaves of the undersized young trees in manufacturing towns expunged all trace of a colourful, seasonal display. In 1904 the *Westminster Review* published the comments of Robert Holland, consulting botanist to the North Lancashire Agricultural Society, on the effects of smoke on flora in Manchester's parks as representative of the conditions in northern factory towns generally:

> Some years ago I had the honour of making an inspection of all the public parks of Manchester on behalf of the Corporation ... I scarcely need say that, going as I did from the fresh green country, I was horrified to see the havoc that was being made. Fine open spaces ... which ought to have been beautiful, and would have been picturesque if well covered with trees, and which should have supplied pleasant recreation grounds for a population that sees far too little of country life and breathes far too little of fresh country air – rendered hideous by the blackness of everything with[in] them – trees stunted, dying, flowers struggling to bloom and sometimes their species scarcely recognisable. It is no exaggeration; and as long as the surrounding chimneys send out volumes of sulphurous acid and of carbon there can be no improvement.[140]

Furthermore, Holland insisted that the destructive influence of the 'smoke nuisance' was clearly discernible over a large proportion of Lancashire and Yorkshire.

With time, the smoke cloud gradually reached beyond the purely local to blight the vegetation of the whole Manchester region and beyond. From the 1840s some contemporaries had been aware that smoke could travel for 'enormous distances', and those that ventured out into the Lancashire country-side in search of nature documented its increasingly damaging effects.[141] Scores of people who lived and worked in the Manchester area spent their few precious leisure hours botanising, gathering specimens of the many different kinds of flora that grew close to the city. In 1859 Leo Grindon estimated that within a fifteen-mile radius of Manchester's Exchange a total of around '1,500 perfectly distinct forms' of vegetation could be found.[142] However, not all species were equally abundant, and Grindon conceded that less than half this number constituted plants that could be encountered generally in the district. Anne Secord has recently pointed out that Lancashire's industrious plant-hunters were so assiduous that they could rapidly strip a locality of its rare flora.[143] Grindon deplored the 'wanton destruction' of plant life by ramblers, but identified air pollution as a greater threat to many species of the region's vegetation.[144] He recorded 73 species of lichen in his *Manchester Flora*, but voiced great concern over their future survival in the region:

> ... the majority of those enumerated are not obtainable nearer than on the high hills beyond Disley, Ramsbottom, Stalybridge, and Rochdale, and even there the quantity has been much lessened of late years, through the cutting down of old woods, and the influx of factory smoke, which appears to be singularly prejudicial to these lovers of pure atmosphere.[145]

Lichens are extremely sensitive to sulphur dioxide and acid rain, and the 'beautiful patches of grey or yellow' which adorned the time-stained walls of old buildings were among the first casualties in the polluted manufacturing towns.[146] Lichens have now started to recolonise Britain's cities, and the classification of lichen flora has proved to be a useful biological indicator of current levels of sulphur dioxide in polluted areas.[147] Moreover, smoke pollution from the early centres of the Industrial Revolution caused 'drastic vegetation change' in the southern Pennine hills. Vast carpets of *Sphagnum* mosses, which had previously dominated in this upland locality, almost completely disappeared due to the extreme acidity of the peat. The cotton grass and bilberry moorland that prevails today replaced them.[148] Ironically, a representative area of the changed moorland of the southern Pennines, its unique features acknowledged to be the result of an 'intense kind of human influence', has been preserved as a grade 1 Site of Special Scientific Interest.[149]

The last decades of the century saw more and more land fall prey to the builder, while the 'smoke nuisance' sullied and destroyed great swathes of vegetation between the expanding industrial towns of south-east Lancashire. In 1882 a local doctor bemoaned the effects of air pollution upon the flora of the entire Manchester region:

> ... fruitful vales where vegetation flourished, roses grew in abundance, and the most delicate flowers thrived, have been changed by the deleterious compounds of coal-smoke into barren deserts. There is now no vegetation, no roses, no

flowers. What once were trees with wide-spreading branches have either disappeared or are represented only by a stunted rotten stump.[150]

Margaret Penn, in her autobiographical novel *Manchester Fourteen Miles*, supports this observation, describing the train journey through the sprawling industrial wilderness that lay between Hollins Green and Manchester thus:

> ... the hideous landscape ... was pock-marked by the belching chimneys of the
> steel works and the soap works and the foundries which bit deep into the flat,
> grey-green country. Now and then, between the soaring chimneys and the
> smoke clouds, she could see a ship's red funnel on the Mersey, the only touch of
> warm colour in the drab scene.[151]

Manchester's flora was unable to withstand the unrelenting, intensifying assault from air pollution. The vagaries of climate, such as heavy rain, high winds, and late frosts, had long been recognised to adversely affect the health and vigour of Manchester's flora. Yet in the past, plant life in the city had always bounced back quickly from natural setbacks. But the synergy of the elements and anthropogenic air pollution dealt a devastating blow to both urban and regional vegetation. Acid rain was one such insidious combination, ruining flora and leading to the widespread acidification of the city's soil and the peat layers of the southern Pennines. Smoke-laden fogs were equally harmful and could have more striking results. When the city was enveloped in a thick sulphurous fog, plants, including those grown under the protection of glass in conservatories and greenhouses, could be denuded of their leaves and flowers in a single night.[152] Damage to vegetation was particularly conspicuous in northern and eastern districts of the city such as Bradford and Ancoats, which lay in the path of the prevailing winds, and where factories were most numerously planted. Grass, for example, could not be cultivated at the Helmet Street playground in Ancoats 'owing to the great acidity of the soil', and this recreation area was as bleak and grimy as the local 'mean streets'.[153] At mid-century the author of *Bradshaw's Guide to Manchester* had argued, 'no one will expect here the gayest side of nature; all her smiles in the fields and lanes are taken off by the whirr and smoke of factories: nature should be seen elsewhere'.[154] By the early years of the twentieth century nature was in full retreat, and *Black's Guide to Manchester* advised those who wished to enjoy beautiful countryside either to cycle out in the direction of Cheshire or to take a train to the Peak District of Derbyshire.[155] However, even Cheshire did not escape the smoke cloud wholly unscathed. Katherine Chorley noted that the leaves of the trees at her home in Alderley Edge, some 15 miles to the south of the city, were 'gradually smirched ... by the creeping grime from Manchester'.[156]

Smoke's destruction of Manchester's flora inevitably had a knock-on effect on its fauna. But, in contrast to the wealth of information that exists concerning the influence of air pollution on vegetation, the Victorians produced a scant amount of detailed evidence regarding the effects of smoke on urban fauna. Perhaps this is because by the nineteenth century large towns and cities housed, besides human beings, only a 'narrow roster of synanthropic species (rats, cats, dogs, cliff-dwelling birds, etc.) and their parasites'.[157] Other than complaints by farmers that livestock was deteriorating 'because good grass does not grow where smoke travels', and irritation that air pollution 'fleeces our sheep of their whiteness', there is little mention of smoke adversely affecting animal life.[158] However, in 1917 William Brend, Lecturer on Forensic Medicine at Charing Cross Hospital, London, recorded the adverse effects of the polluted metropolitan atmosphere on captive zoo animals:

> The death-rate among animals surrounded by unnatural conditions is very high, and often their young can only be reared by taking the utmost precautions. London reads with regret the fate of litter after litter of the cubs of 'Barbara', the polar bear in Regent's Park, which live at the most for a few weeks and then die from pneumonia. The higher apes suffer severely from tuberculosis, and nearly all the mammalian cubs develop rickets.[159]

And by the early years of the twentieth century it was believed that a lack of sunlight and clean air made dairy cattle more susceptible to bovine tuberculosis.[160]

With regard to avifauna contemporaries observed that many species of birds were forced out of urban areas by the 'smoke nuisance'. The Lancashire ornithologist F.S. Mitchell noted that human activity endangered bird life in many ways, but that it was coal smoke above all else that caused the greatest harm:

> ... birds are driven away by the extension of buildings, and by the conversion of a rural into an urban locality ... [but] the main causes are, not so much simply the presence of more people and greater disturbance by them, as the destruction of natural food, and the loss of protective foliage from the vitiated atmosphere.[161]

The gradual loss of mature trees, shrubs, and plants considerably reduced habitat and food supply for resident birds, and in turn the number of species to be met with in the city began to fall dramatically. For more than three decades John Plant of Salford kept meticulous notes of the birds he observed at Peel Park. According to Plant, of Lancashire's 259 bird species a highly creditable 71 varieties visited the park between 1850–60, with 34 species breeding and producing offspring. Between the years 1870–75 he counted just 19 species, with only 8 different varieties rearing young. In 1882, the final year for which figures are given, Plant personally observed a meagre 5 types of bird at Peel Park and only the house sparrow and starling were to be found nesting there.[162] In addition, kestrels and other raptorial birds, which might have preyed on burgeoning urban pests such as mice and starlings, are dependent on their keen eyesight to secure their next meal. High levels of smoke pollution obscured the hunted from the hunters, and these hovering birds were rarely seen in central Manchester before the 1950s.[163]

As was the case with urban flora coal smoke not only reduced the diversity of species and reproductive success, it also affected the outward appearance of the hardy bird species that managed to maintain a presence in the centres of industrial towns. The black adhesive film of soot that besmirched urban vegetation attached itself to the plumage of the town bird, leading to unflattering comparisons with their country cousins. For example, in 1896 Julius Cohen commented,

> The very birds of the air are tarred with this universal tar brush ... a Leeds magpie, shot near Stainbeck Lane, ... bears evident signs of his town residence. Not only are the white feathers badly discoloured, but there is a striking absence of the gloss and beautiful iridescence of the black ones, visible in [the] country magpie.[164]

Moreover, a working class child from London, confronted by an unsoiled avifauna when taking a rare holiday in the countryside, remarked, 'The birds are not like ourn they are light brown.' [165] The murk of the Victorian city, however, proved beneficial for at least one urban species – the melanic form of the peppered moth, *Biston betularia*. As smoke killed the lichens that provided protective cover for the pale coloured typicals, blackening tree trunks and the walls of buildings, the dark peppered moth became the predominant type in Manchester. Initially few in numbers, the first melanic specimens were not caught in the city until 1848. By 1895 the black type was prevalent, with the white form making up only two per cent of Manchester's peppered moth population. Only after smoke control began in earnest in 1952, and buildings were cleaned and lichens started to return, was there a resurgence of the typical variety in the heart of the city.[166] The 'smoke nuisance' affected every living thing in nineteenth century Manchester, substantially changing and depleting its flora and avifauna, and turning those organisms which endured a depressing shade of black. Furthermore, the polluted quality of the atmosphere posed a very similar set of problems with regard to the built environment.

Smoke and architecture

Between 1840 and 1860 Manchester's architecture was widely regarded as the most original and advanced in Britain.[167] The principal foundations for Manchester's ascendancy lay in the adoption of new construction techniques, in the utilisation of materials such as iron and terracotta, and in its architects' adaptation of the elegant *palazzo* architectural style of Renaissance Italy. Charles Barry first introduced this style of architecture into the city in 1837 in his design for Manchester's Athenaeum, which was based on Raphael's *Palazzo Pandolfini* in Florence. The Athenaeum's humanistic symmetry, plain surfaces, large, closely spaced windows, and bold features provided the blueprint for many of the city's great warehouses, which were so exciting the interest and imagination of contemporaries.[168] In 1844 Benjamin Disraeli declared that 'Manchester is the most wonderful city of modern times!', while the city's 'commercial palaces' were eulogised by *Bradshaw's Guide to Manchester* in the following terms:

> Here are structures fit for kings, and which many a monarch might well envy. There are some eight or ten sovereign princes in Germany whose entire revenue would not pay the cost of these warehouses. The industrial and scientific energy that has reared them is an honour to our country, and speaks well for the future of Manchester. The artistic display is all but equal to the noble enterprise which gave them being. They are, indeed, the most splendid adornment of this city, and really monumental, whether we regard their splendour, their proportions, or their durability.[169]

Beauty and good taste, however, were not the only factors in facilitating the widespread use of this grand architectural style. Victorian architects, as Michael Brooks has observed, worried constantly about smoke, 'They thought of it as the source of a practical problem in design, one that they could not solve.' [170] The suitability of the *palazzo* style to the polluted environmental conditions of the northern industrial towns made it an inspired and popular choice, and one that undoubtedly helped to mitigate the worst effects of the 'smoke nuisance'. Large windows, and plenty of them, ensured the maximum admittance of all available natural light in the gloomy industrial city. The austere exterior and restrained ornamentation of these buildings were designed not only to express the sober and dignified characters of the northern 'merchant princes' who commissioned them, but also to hinder the inevitable accumulation of soot and grime. Moreover, the city's architects were as careful in their choice of building materials as were its gardeners in their selection of flora. Through their characteristic combination of red brick and stone trim, and their use of terracotta, Manchester's architects added a welcome dash of warmth and colour to the drab city streets. But, at the same time, they were well aware that these materials were more resistant to the viscid smoke than, for example, rough, open-textured limestone.[171] The smoother surfaces of the brick walls, often glazed, made it hard for soot to adhere, and bricks were also easier to clean.[172] Despite the difficulties caused by coal smoke, mid-Victorian Manchester had acquired the reputation for having fine public and commercial buildings lining its principal streets.

In the 1850s there was a movement towards greater richness of detail and of more ornamental carving in the *palazzo* style of architecture, and Manchester's architects again led the way. Edward Walters' warehouse of 1851, built for James Brown and Son at Portland Street, was the most prominent and perhaps the most impressive example of this architectural style. It combined a different treatment of the elaborately carved windows of each of its four storeys, with the edifice ultimately crowned by the great urns of its high balustrade. Again, the city's smoke affected fundamentally the way in which architects such as Walters could practise their art. He realised that in Manchester only the boldest of ornamentation could be expected to remain visible under a patina of soot and against the backdrop of dark, smoke-laden skies. Walters deliberately compensated for the city's polluted atmospheric conditions by increasing the coarseness of his mouldings, and making the relief of his carved details sharp and overscaled.[173] Moreover, Victorian architects, inspired by the abundance of coloured materials used in constructing venerated Renaissance and Gothic buildings, found that smoke constrained their use dramatically in the industrial towns of Britain. In 1861 the journal *Building News* voiced concerns that air pollution was negating the use of architectural colour, while a Manchester architect later wrote:

> What splendid opportunities those old builders and architects had of exercising their imaginations, revelling in the additional charm beyond that merely of form, which the use of colour gives ... What a glorious wealth of colour there is in the old front of St. Mark's at Venice! with its coloured marbles, imperishable frescoes, bronzes, old gold and masonry, mellowed, enriched, and toned by the hand of time almost into an architectural dream ... *Because* of this soot and dirt, our clients won't let us make use of materials and modes of decoration which we have at hand. We have granites of various and beautiful colours, rich Devonshire and Irish marbles; we have beautiful clays capable of being moulded and modelled to any shape, coloured to any tint and glazed; we have beautiful woods, glass, tiling, mosaic, skilfully worked metal, &c., &c., and *we know how to use them*; we can design beautiful ceilings and decorate them, as well as our walls, appropriately, but we are everywhere confronted with the demon *smoke*, who limits our artistic aims within the prosaic grounds of practical use ...[174]

But, if the smoky atmosphere meant the prospects of Manchester's built environment being enriched by colour other than red brick and terracotta, or slowly enhanced by the hand of time, were slim, then there were other ways in which enterprising Victorian architects could express their artistic talents.

After mid-century, inspired by the works of John Ruskin, architects in the prosperous northern towns began to design buildings in the Gothic style. The symmetry of the *palazzo* style had announced the might of the human intellect, and celebrated human power and mastery over nature. Rejecting such formal perfection, the Gothic style centred on evoking natural forms and 'in effect confess[ed] the weakness of man and acknowledg[ed] that only God is perfect'.[175] The architects of the Gothic Revival honoured the glory of God's creation – nature – in their edifices. It was believed that good Gothic architecture could restore something of nature to the barren industrial city, as Ruskin had asserted:

> We are forced, for the sake of accumulating our power and knowledge, to live in cities: but such an advantage as we have in association with each other is in great part counter-balanced by our loss of fellowship with Nature. We cannot all have our gardens now, nor our pleasant fields to meditate in at eventide. Then the function of our architecture is, as far as may be, to replace these; to tell us about Nature; to possess us with memories of her quietness; to be solemn and and full of tenderness, like her, and rich in portraitures of her; full of delicate imagery of the flowers we can no more gather, and of the living creatures now far away from us in their own solitude.[176]

The underlying principle of the Gothic style was to promote harmony between natural environment, architecture and human sensibility. Indeed, the edifice must actually appear to 'grow out of the land'.[177] The luxuriant ornamental work on Gothic buildings depicted natural foliage, blossoms, fruits, and myriad members of the animal kingdom. Gothic arcades and spires were designed to elicit associations with primal forests and mountain peaks. Manchester, its streets lacking almost all vestiges of vegetation, was forced to rely heavily on its architecture to bring some semblance of natural beauty back to the polluted city. The leading architect of the Gothic Revival in mid-Victorian Manchester was Alfred Waterhouse, whose design for the city's Assize Courts in 1859 delighted Ruskin who thought it was 'much beyond everything yet done in England on my principles'.[178] Its most notable feature was the fine ceiling of the Great Hall, which had the quality of a high forest canopy. In 1861 Waterhouse consolidated his burgeoning reputation by building the distinctive Royal Insurance Building in King Street, which boasted a trio of storks perched quizzically on the arches of its dormer windows. Waterhouse's friend and great contemporary rival, Thomas Worthington, incorporated strawberry leaves and figures denoting Spring, Summer, Autumn and Winter into his design for the 'medieval ciborium' constructed to house the marble statue of Prince Albert in 1867.[179] A decade of monumental building work in the Gothic style was capped in April 1868 by Waterhouse's winning design for Manchester's new Town Hall, the most striking element of which was a magnificent spire reaching almost 300 feet in height. Opened in 1877, Waterhouse's lavishly decorated neo-Gothic masterpiece, completed at a cost of £1 million, was perhaps the finest town hall in the country.[180] When the building work on the Town Hall was largely completed, Thomas Worthington told his colleagues in the Manchester Society of Architects,

Manchester is acquiring the reputation of a town of some architectural character; it is the inland metropolis of the North – the Florence, if I may describe it, of the nineteenth century; it has developed a style of architecture which we may largely call our own and in which we may take a not unnatural pride.[181]

However, the 'abnormal or accidental constituents of the air of towns' – acid rain and tarry, acid-laden soot – were recognised by contemporaries to be causing the decay of building materials at a significantly faster rate than the erosion which occurred naturally through weathering by wind and precipitation in country areas. As early as 1859 Robert Angus Smith had detailed some of the deleterious effects of smoke thus:

It has often been observed that the stones and bricks of buildings, especially under projecting parts, crumble more readily in large towns where much coal is burnt than elsewhere ... I was led to attribute this effect to the slow but constant action of the acid rain ... it is not to be expected that calcareous substances will resist it long, and one of the greatest evils in old buildings in Manchester is the deterioration of the mortar. It generally swells out, becomes very porous, and falls to pieces on the slightest touch.[182]

The architecture of the city dubbed the 'Florence of the North' soon began to darken, and its elaborately carved stonework to perish, due to the corrosive action of the sulphurous smoke cloud. The lofty aspirations of Manchester's manufacturing and commercial middle class had found their expression in innovative, monumental architecture and statuary.[183] But buildings that were designed to reflect the sophistication, power, and status of the city's 'urban aristocracy' *ad infinitum* were rapidly despoiled and their artistic value nullified by the 'smoke nuisance'. In 1859, due to the city's 'general smokiness', the *Builder* lamented, 'As to the architecture in the streets of Manchester, there are few towns we imagine, where the realization of effect bears so trifling a proportion to the endeavour, and, indeed, to the merit of the art.'[184] By the last quarter of the century, despite the erection of numerous architectural works of art, Disraeli's 'wonderful city' was widely considered to be 'uglier than nearly all other towns'.[185] In 1887 a correspondent to the *Manchester Guardian* complained that, due to the damaging effects of smoke, 'Manchester in attractiveness is far behind her sister towns of equal or greater magnitude ... our own unloveliness is great compared with such cities as London, Liverpool, Dublin, Edinburgh, Glasgow.'[186] The rich detail of Worthington's Albert Memorial had by this time become indecipherable beneath a thick deposit of carbon, while the statue of the Prince Consort was thought to resemble 'a representative of one of the dark races'.[187] The fate of Manchester's inventive and varied architecture may be summed up by the experience of the ultimate Gothic monument to civic pride and commerce – the Town Hall. In a short space of time its outside walls were soiled and streaked a sombre black and its ostentatious ornamentation became 'mere traps for soot and grime'. Thomas Horsfall was moved to question the wisdom of spending a million pounds on a town hall when 'most of its architectural effect would be lost because [it was] ruined by soot and made nearly invisible by smoke'.[188] In the smoke-filled industrial city the naturalistic decoration and forms of the Gothic style of architecture fared little better than the natural phenomena it was designed to replace. Hopes that strongly carved, lifelike representations of flora and fauna, or soaring spires, could bring back the essence of nature to the people of Manchester were promptly dashed by the 'smoke nuisance'. If architecture conveys and celebrates the values of a society, then the eclectic buildings of Victorian Manchester, uniformly clad in a velvety black coating of soot, accurately reflected the importance of coal and steam power to contemporary urbanites.

The urban smoke cycle

If air pollution perturbed natural cycles in the city, at the same time the 'smoke nuisance' imposed its own sustained rhythms on urban life that fluctuated daily, weekly and seasonally. Changing climatic conditions meant that day to day concentrations of smoke pollution varied widely in urban-industrial areas. However, levels of smoke pollution also ebbed and flowed in line with relatively settled patterns of coal consumption in factories and private houses. On a weekday the smoke cloud rapidly expanded between the hours of 6 and 7 a.m. in Manchester, as domestic fires were rekindled and the chimneys of manufacturing industry – largely dormant at night – belched back into life. The smoke was at its thickest around 11 a.m., with a subordinate peak at 5 p.m.; periods which coincided with the preparation of midday and evening meals.[189] Pollution gradually waned after this point, as mills and factories wound down and domestic fires were allowed to burn low at the end of the day. The smoke cloud contracted to its lowest ebb between the hours of 10 p.m. and 4 a.m., although it rarely happened that there was no smoke at all during the 24 hours.[190] On Sundays the peaks of the smoke cycle occurred one or two hours later than on a weekday, as the workers liked to sleep late on their day off. Generally speaking, Sunday was the day on which Manchester's air was least polluted as the city's factories were closed. However, this rule of thumb did not hold true for many non-industrial towns and cities, nor for the crowded suburbs of industrial towns, where smoke pollution on Sundays was thought to exceed that of the average weekday.[191] The quantity of extra fuel required for cooking the main meal of the week, Sunday dinner, was held to be the major culprit. A prodigious amount of roasting, grilling, baking, frying and boiling produced a surfeit of smoke from many thousands of domestic chimney pots. The *Public Health Engineer*, for example, reported that the urban smoke haze was at its worst on Sundays 'when extra cooking took place over the widest area'.[192]

Illustration 3. Domestic Chimneys at Manchester.

Source: Manchester Central Reference Library.

The everyday habits of the people determined the pattern of the weekly smoke cycle, with Monday and Tuesday being the days of greatest fuel consumption. If Sunday was Manchester's clearest day, Monday, the day on which industry resumed its operations, was the smokiest day of the week. For early on a Monday morning most working-class women lit substantial fires to procure the large quantities of hot water necessary to carry out the weekly wash.[193] The washing, drying, and particularly the ironing of clothes frequently carried over into Tuesday making it the second most polluted day of the week.[194] In the city's damp and severely polluted atmosphere it was sometimes impossible to hang washing outdoors to dry.[195] Clothes had often to be dried and aired indoors, and, in addition, a good-sized coal fire was needed to heat the common flatiron before the arduous task of ironing could begin. The graph shown below in Figure 1, constructed by William Thomson of the Manchester and Salford Sanitary Association, shows the average daily intensity of smoke pollution in Manchester for the nine month period between September 1914 and May 1915.

The steep downward curve of the graph from mid-Tuesday to Sunday can be explained not only by the orgy of laundry work undertaken during the first part of the week, but also by the fact that regular stoppages at the city's mills and factories due to short-time working usually occurred at the week's end.[196] And Saturday, of course, saw less industrial activity in any case, as it was a half-day holiday. The dislocation of the cotton trade caused by the outbreak of the First World War must have slightly exaggerated the curve of the graph.[197] But

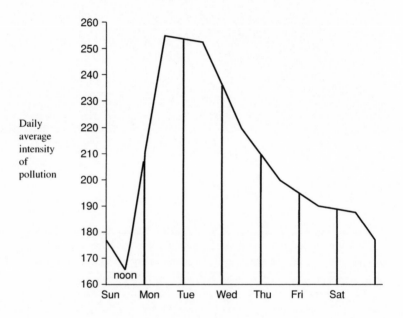

Figure 1. The Urban Smoke Cycle.

Source: William Thomson, *Annual Report of the Manchester and Salford Sanitary Association*, 1915. The numerical values reflect the dark colours of the impressions made by coal smoke on the paper filters of Thomson's apparatus.

unemployment and short-time working were common features of life in Manchester, and, even allowing for this extraordinary disturbance of trade, the findings of Thomson's investigation are likely to represent a fairly reliable picture of the city's weekly smoke cycle for much of the Victorian period too.

However, Thomson's daily averages mask the seasonal variations in the density of the smoke cloud, which were also evident to contemporaries. Smoke billowing forth from industrial chimneys followed the fluctuations of the trade cycle, while smoke issuing from domestic chimneys waxed and waned with the passing of the seasons. During the cold winter months heat was the prime consideration and coal fires were kept burning all day, often in several rooms at the same time in middle-class homes. Moreover, the radiant form of heat emitted by a bright coal fire was widely believed to be an excellent antidote for a gloomy winter's day. For example, the influential architect C.J. Richardson opined, 'the chief point of attraction in the English dwellings, during winter's wet, cold, and fog, is centred in the fireplace ... [it] suits our climate; it is cheerful and attractive'.[198] Warming the house was an important function of the coal fire in spring and autumn too, and every cold snap increased the volume of air pollution in the city. But in summer a lone, modest fire for cooking, and obtaining the small amounts of hot water necessary for bathing and washing dishes, was sufficient to meet a household's needs for most of the week. Nonetheless, during the summer months the 'smoke nuisance' maintained a permanent haze above urban-industrial areas, and in 1855 the *Times* described London's air pollution thus:

> Smoke we have always with us. If we look out on a fine summer's day through the louvre boards at the top of the Crystal Palace for a view of the great metropolis, we naturally exclaim, 'I see it; there is the smoke;' indeed, any picture of London without its dim canopy of soot would be as unrecognizable as would a portrait of Pope, Hogarth, or Cowper without their well-known headgear.[199]

Even so, the diminution of smoke from private houses in summer, combined with brilliant sunshine, did produce an appreciable difference in the quality of London's atmosphere. Ernest Hamilton gave the following account of the metropolis, basking in the summer sunshine, to the *Pall Mall Magazine* in 1894:

> London in June is hardly recognisable as the same place where six months before we were coughing and wheezing and groping our grimy way through the gaslit streets. In June the trees are in the full zenith of their short-lived verdure, the young grass fresh and green, the parks bright with flowers, and the exhalations of domestic chimneys have ceased for a time to obscure the heavens. In short, everything looks its best and brightest, and only the houses stand as gloomy, silent witnesses that the truce with the powers of darkness is only temporary ... So it is year after year. We grumble in winter, and we forget in summer.[200]

At Manchester scientific investigations concerning air pollution's effects on light in the city and its suburbs, carried out by the Air Analysis Committee of the Manchester Field-Naturalists and Archaeologists' Society, supports impressionistic evidence concerning the seasonal variation of the smoke cycle. The committee discovered that in both 1891 and 1892 the atmosphere was at its clearest in September, while November, December, and January were the city's darkest months.[201] Their experiments also revealed that in winter the city's most densely populated residential districts, such as Hulme, received on average only half the light of nearby suburbs like Didsbury.[202] However, even during the brightest weather the city's ever-changing smoke cloud was found to be extremely effective in blocking out sunlight. Statistical evidence of average yearly sunshine records for the years 1905–09 unsurprisingly shows Manchester trailing far behind Bournemouth, Cambridge and other residential towns. Furthermore, the city languished at the foot of a 'sunshine league table' of manufacturing towns that included Birmingham, Bradford, Glasgow, Leeds and Sheffield, as Manchester received only 25 per cent of all possible annual sunlight.[203] The meteorologist Frederick Brodie recorded a very similar yearly average for bright sunshine at Manchester between the years 1881–1910 of 26 per cent of the available duration. But, more importantly, he disclosed that Manchester's average fell markedly to just 11 per cent of what may be termed 'expectation of sunshine' during the months December to February, when the demand for a warm, bright coal fire was at its greatest.[204] The smoke cloud, then, was a durable shifting mass, constantly altering its bulk and rhythms under the variable influence of climatic conditions and human action. Industrial smoke was the underlying wellspring of the city's uneven smoke cycle, while the periodic habits and routines of the people played an unspectacular but crucial role in dictating the ebbs and flows of air pollution in Manchester. But what effects did the smoke-filled skies have on the day to day lives of Manchester's citizens?

Living with smoke

Monday or Sunday, winter or summer, the domestic hearth was the hub around which family life revolved. The coal fire provided warmth and light, hot water for bathing and washing clothes, as well as cooked meals. In addition, it also functioned as an incinerator for most household refuse, provided much-needed ventilation of rooms, and furnished ashes for use in Manchester's many thousands of dry conservancy pail privies. In return for these vital services, this household god required continual devotion and attention from those who worshipped at its altar. Every day the same ritual was performed: the grate had to be cleaned, the hearth swept, and the coal scuttle filled, before the business of laying the fire could begin.[205] Once the fire was burning it needed to be fed fresh supplies of fuel at regular intervals or it would expire. This could be hard, dirty work, as the diaries of the Victorian housemaid Hannah Cullwick reveals. In one of Cullwick's situations she was expected to clean and lay no fewer than five fireplaces before breakfast, a demanding feat which also involved removing the ashes and carrying heavy coal scuttles up and down stairs.[206] It was women – as housewives or domestic servants – rather than men who sacrificed much of their time and energy on tending and servicing the domestic hearth.[207] Similarly, it was women who had to cope with the Sisyphean task of keeping homes clean in smoky industrial towns.

In addition to holding down paid employment, washing, cooking, and child rearing, cleaning the home was an important, if undervalued, part of everyday life for the Victorian housewife. There were notable exceptions, as middle and upper class 'ladies' employed armies of domestic servants to undertake such 'degrading' work, while many lower-class working women paid for help with the household chores too.[208] Even so, house cleaning was a time-consuming activity that locked most women – and a few males – into a never-ending round of drudgery.[209] The all-pervasive dirt and smuts from the soft, bituminous coal burnt in the countless furnaces and hearths of Lancashire made a stubborn, unrelenting foe. Nevertheless, on numerous occasions contemporary onlookers, such as Samuel Bamford and Angus Reach, described the interiors of workers' cottages in Manchester and neighbouring factory towns as scrupulously neat and spotlessly clean.[210] But as the smoke cloud deepened by degrees, the wearisome daily battle against dirt became more and more difficult to win. In 1882, the anonymous female author of an article against smoke stated,

> We have a great respect for the average British working-woman, and believe her to be a good deal more industrious than the average British workman; but to keep house and children and self clean in a back slum of a manufacturing town is a feat beyond human power.[211]

The soot and grime that destroyed vegetation and blackened the exterior surfaces of buildings also filtered through the narrowest cracks and fissures to soil everything within the home. Walls and ceilings were smeared a dingy black, fittings and furnishings were covered in black dust, and windows seldom remained transparent for more than a few days. Legions of house-proud women dusted, scrubbed, and polished, often for hours at a time, to try to keep what they possessed clean.[212] Indeed, Robert Roberts observed that at Salford many women did succeed in holding dirt at bay, but only at a price:

> Some houses sparkled. Few who were young then will forget the great Friday night scouring ritual in which all the females of a house took part. (Dance halls closed on Friday evenings for lack of girls.) Women wore their lives away ...

scrubbing wood and stone, polishing furniture and fire-irons. There were housewives who finally lost real interest in anything save dirt removing. Almost every hour of the week they devoted to cleaning and re-cleaning the same objects so that their family, drilled into slavish tidiness, could sit in state, newspaper covers removed, for a few hours each Sunday evening. On Monday purification began all over again. Two of these compulsives left us for the 'lunatic asylum', one of them, I remember vividly, passing with a man in uniform through a group of us watching children to a van, still washing her hands like a poor Lady MacBeth.[213]

On the other side of the coin, Rollo Russell claimed that because of London's heavy smoke pollution some women had ceased to care about cleaning: 'many who have been good housewives in the country ... give up the attempt to keep their houses clean'.[214] And at Manchester vivid contemporary accounts of squalid living conditions indicate that considerable numbers of the poor had discontinued the futile daily struggle against filth and grime.[215] On balance, although homes never remained clean for very long, it seems likely that most working-class women did at least attempt to keep their homes free of dirt. What is more, by the end of the nineteenth century the fastidious, beleaguered Lancashire housewife had become a by-word for the type of woman who was a 'slave to cleanliness'.[216] The enormous amount of time and effort invested in performing housework draws attention to a strong, commonplace desire to maintain a relatively clean and comfortable domestic environment: and not simply for the sake of 'keeping up appearances'.

In an age of smoke the ephemeral sensory pleasures associated with open, verdant pre-industrial landscapes – of looking up at shifting cloudscapes or enjoying the fragrance of a flower – were muted or denied altogether to the urban masses. Surrounded as they were by the sombre tones, pungent odours and annoying grime of the perpetual smoke haze, the Victorians sought refuge in their oft-scrubbed homes. With curtains closed against the harsh environmental conditions, and a glowing coal fire creating 'a summer i'th chimdy', large numbers of city-dwellers escaped the ugliness of the outside world by retreating into a snug domesticity.[217] For those whose houses were too dirty and uncomfortable to warrant spending much time inside them, the warm, brightly-lit public house or gin palace was a popular surrogate. The darkness that enveloped the northern factory towns, and their barren, smoke-filled streets, played a fundamental, if understated, role in keeping people indoors. Yi-Fu Tuan has called attention to the development of 'cozy' indoors recreational activities, such as chamber concerts and afternoon teas, as an 'adaptive response' to industrial pollution in Victorian Britain.[218] The heavily polluted atmosphere, along with the damp north-western climate, undoubtedly put paid to many *alfresco* social activities and helped to produce an insular, stay-at-home urban populace, as Julius Cohen pointed out in 1896:

> ... if we could sit after the fashion of the Continentals, and drink coffee and smoke under the trees outside club or restaurant, we should begin to feel dissatisfaction with some of the hideous surroundings which we now tolerate ... But the gloom ... drives us ... within doors, where we strive to compensate for the ugliness and discomfort of outside, by the enhanced cheerfulness and attractiveness of our homes. I believe that in consequence of this, we have reached a standard of domestic comfort which is unknown in any other country [and] ... Any pride we may have in our surroundings rarely extends beyond our own four walls ...[219]

Deprived of sunlight and natural colour, the inhabitants of the factory towns did their best to create for themselves an intimate and stimulating home environment. In *Mary Barton*, Elizabeth Gaskell carefully detailed the inviting interior of a 'respectable' working-class home. Included among its welcoming attributes were whitewashed walls, a 'gay-coloured piece of oil-cloth', and a jumble of furniture, crockery, and glass, all set off to good advantage by the firelight which 'danced merrily' about the room.[220] Gaskell also alluded to the 'childlike' working-class taste for garishly coloured objects, such as the 'bright green japanned tea-tray, having a couple of scarlet lovers embracing in the middle'.[221] The rising wages of the urban masses enabled them to purchase novel and inexpensive consumer goods of all sorts, which were then proudly displayed to brighten up the home. Cheap prints of sentimental or country scenes bedecked living-room walls, and fire irons gleamed while every available surface was covered with gaudy knick-knacks.[222] The cluster of ornaments on show above the fireplace at the home of Robert Roberts included: two massive dogs (with pup), a hollow bust of William Gladstone, a pair of gondolas, and a bowl of china cherries.[223] One of the most influential Victorian commentators on home decoration, Charles Eastlake, reported in shocked tones that he had even caught sight of a floorcloth designed to 'represent the spots on a leopard's skin'.[224] With regard to interior decoration, two other things would also have struck the contemporary observer. Firstly, the fine displays of hardy house plants such as the common fuchsia, geranium, and the ubiquitous aspidistra; and, secondly, the way in which ornate floral designs wound their way across a host of domestic artefacts, from carpets, curtains, and carved furniture to wallpapers, crockery, and even coal scuttles.[225] Both trends reveal a yearning for the greenery that no longer existed out of doors in many urban-industrial areas. No corner of the home was left bare and the Victorians have been roundly criticised for their dubious taste and discordant interiors.[226] But, in the absence of sufficient external stimuli, their predilection for packing the home cram-full of plants and colourful ornamental objects becomes more readily intelligible.

At the same time, however, the gloomy atmosphere that hung over the industrial towns was mirrored by a growing utilisation of practical, dark coloured materials. For example, paints, furniture, and clothes were all deliberately chosen with the smoky air in mind. As one consumer put it, 'When making [a] purchase, [the] question is – will it show the dirt soon?'[227] The light, fresh colour schemes favoured in Regency and early Victorian rooms were gradually discarded for more functional, dull shades of brown, red, and bottle green that did not show sooty streaks so readily.[228] Nonetheless, despite the widespread adoption of a functional range of appropriately drab colours, it was alleged that the constant need for repainting made Manchester a 'paradise' for painters and decorators.[229] At around mid-century furniture began to be manufactured in the darkest woods available, mahogany, bog oak, black walnut, and rosewood, with artificially blackened oak also becoming a very popular material.[230] Moreover, high levels of air pollution in the factory towns made it extremely impractical to wear light coloured clothing, and both men and women were compelled to dress in dark apparel.[231] 'Manchester suits for Manchester soots' was the aptly barbed slogan of one enterprising tailor in the city.[232] And laundries, like painting and decorating firms, were reputed to be flourishing in the city, as curtains and clothing had to be washed far more frequently than in non-industrial towns.[233] Margaret Penn observed that clothes which could have been worn 'clean and stiff for a month of Sundays' at Hollins Green were 'not fit for putting on a second time' after just a single encounter with 'the black dirt of Manchester'.[234] The 'smoke nuisance', then, touched every aspect of domestic life, powerfully influencing a multitude of common everyday decisions and activities. Air pollution greatly increased the burden of washing and housework for women, stimulated a taste for comfort and ornamental decoration, and encouraged people to spend far more time indoors. Furthermore, smoke affected the quality of life for people living in urban-industrial areas in a more fundamental way, by seriously damaging their health.

Smoke and health

In the textile districts of Victorian Lancashire and Yorkshire the numerous advertisements for 'lung tonics' and cough lozenges that peppered popular dialect journals and almanacs, such as *Bill O' Jack's Lancashire Monthly* or John Hartley's *Clock Almanac*, bear witness to the prolonged suffering of untold thousands who coughed and wheezed away their days.[235] The protracted and mundane nature of respiratory conditions, with victims often taking many years to die, helped to obscure their importance as major killers in urban areas. Ellen

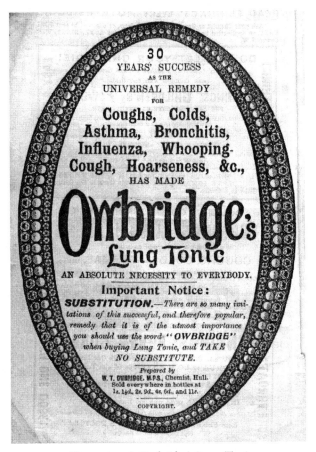

Illustration 4. Owbridge's Lung Tonic.

Source: John Hartley's *Original Clock Almanac*, 1905.

Ross has commented on 'the success of an entire culture' in 'not seeing' chronic illnesses that brought considerable pain and discomfort, but which were not thought to be immediately life threatening.[236] Similarly, spiralling mortality rates from respiratory diseases in nineteenth century Britain have, with the notable exception of pulmonary tuberculosis, thus far received little attention from historians of medicine and public health. Historians have mainly concentrated on analysing responses to the dramatic outbreaks of epidemic 'filth' diseases associated with insanitary urban conditions.[237] In focusing on preventative campaigns against the classic Victorian killer diseases of cholera, typhus, and typhoid, the insidious and 'increasingly lethal' bronchitis group of diseases has for the most part been ignored.[238] Simon Szreter has recently drawn attention to the growing number of fatalities from this quarter in England and Wales during the second half of the nineteenth century:

> ... the composite airborne category, 'bronchitis, pneumonia and influenza', ... was the second most important cause of death in 1848–54, accounting for 10.25 per cent of all deaths. It actually registered a very considerable *absolute increase* in mortality of well over 20 per cent down to 1901. By the turn of the century this category was clearly the most important single killer, contributing over 16 per cent of all deaths, a greater proportion of the total than respiratory TB had represented in the mid-nineteenth century.[239]

The upturn in mortality from the bronchitis group of respiratory diseases was in sharp contrast to a considerable reduction in deaths from infectious 'filth' diseases such as cholera, typhus, and typhoid after mid-century. Indeed, after the final epidemic of 1866–67 deaths from cholera dropped to such an insignificant level that they hardly featured at all in the Registrar-General's annual reports. By the 1870s bronchitis had unobtrusively become the commonest cause of death in England's factory towns, consistently killing between 50,000 and 70,000 people per annum, as Figure 2 illustrates.

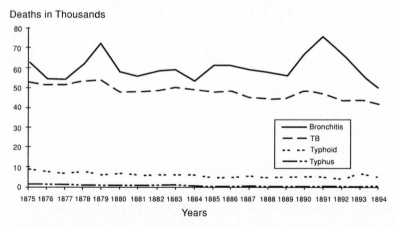

Figure 2. Deaths from Specific Causes in England, 1875–94

Source: Registrar-General's Annual Reports. The sharp peaks in bronchitis mortality were attributed to exceptionally cold weather in 1879 and the related effects of pandemic influenza in 1889–92.

Improvements in sanitation systems, water supplies, nutrition, housing, childcare practices, and health education saw Britain's general death rate falling from an average of 22.2 deaths per thousand living during the decade 1851–60 to 18.2 per thousand in the decade 1891–1900.[240] Manchester's mortality rates also declined during the same period, from 31.5 per thousand in 1851–60 to 25.5 per thousand in 1891–1900.[241] Alan Kidd has noted that almost '88% of Manchester's total reduction in mortality between 1850 and 1900 came from a decline in infectious diseases'.[242] Contemporary concerns about the health risks posed by air pollution increased as the threat from 'dirt' diseases subsided, but the campaign against the 'smoke nuisance' never gained the same momentum or sense of urgency as that waged against filth. Nonetheless, sanitary reformers constantly highlighted the unhealthy nature of a heavily polluted town atmosphere in comparison with more wholesome country air. For example, in 1881 John Tatham, Medical Officer of Health for Salford, accounted for the better health of the country folk of rural Cheshire thus:

... in Salford 598 people in every 100,000 of the population die annually of lung complaints, as compared with only 334 in Mid-Cheshire, a combined sanitary district not more than a few miles distant, and containing at the last census a population of 136,000 people. The conditions of life in this district are not superior to those in Salford, with the one exception that the atmosphere is less contaminated by smoke. The people generally are not more prosperous or better fed, and the climate is certainly not warmer than that of our own town, so that the extreme difference in mortality from respiratory disease may be assumed to be mainly if not entirely due to the smoke nuisance.[243]

Anti-smoke activists often used statistical data to link air pollution to the growing incidence of chronic respiratory diseases in urban-industrial areas. One such report showed that the average death rate from respiratory diseases between the years 1868–73 was just 2.27 per thousand in rural Westmoreland, 3.54 per thousand for the whole of England and Wales, while for Salford it was 5.12, and for Manchester 6.10 per thousand. In 1874 the death rate from these causes was 7.70 per thousand in Manchester, leading the report's author to conclude, 'Manchester suffers more from diseases of respiratory organs than any town or city in England'.[244] In 1878 Dr Arthur Ransome, a prominent member of the Manchester and Salford Sanitary Association, calculated that in the twin cities during the decade 1865–74 some 22,191 people had died from respiratory illness and a further 12,343 from pulmonary tuberculosis.[245] And if contemporaries accentuated the contrast between city and country fatalities, spatial variations with regard to death rates within urban areas were also brought to notice. In the late 1880s, at a time when improvements in Manchester's overall death rates were first becoming evident, an investigation undertaken by Dr John

Thresh exposed the appalling mortality of No.1 sanitary district, Ancoats. Thresh found that in mainly residential working-class areas with similar (or higher) population densities per acre, death rates were significantly lower than in industrial Ancoats. For example, he noted that 'whilst Ancoats No.1 has a population density of 201 per acre, and a death-rate of 28.7 per 1,000, Hulme No.8, with a population density of 211 per acre has a death-rate of only 18.8 per 1,000 – an astonishing difference truly'.[246] What is more, after tracking down many of Ancoats' sick and dying to Crumpsall Workhouse and other hospitals outside their usual area of residence, Thresh pronounced: 'The true death-rate of the whole district is seen to be over 50 per thousand, whilst in the courts it exceeds 80, and in one street actually has exceeded 90.' Maria Street, hemmed in on three sides by Ancoats' huge mephitic mills, recorded a death rate of 91.6 per thousand in 1887.[247] By the end of the century Ancoats was recognised to be Manchester's mortality black spot, with its death rates from respiratory diseases being some three times higher than in other districts of the city.[248]

The young, the elderly, and those already in poor health were most susceptible to the deleterious influence of air pollution. Age-specific patterns with regard to deaths from respiratory disease were stressed by a *Lancet* report, which analysed mortality in Islington and St. Pancras between the years 1870–97. The report's author, W.T. Russell, Chief Statistical Clerk at the National Institute for Medical Research, found that in the two boroughs in question children under five and adults over 55 constituted some '70 per cent of the total deaths from respiratory diseases'.[249] Moreover, that freezing, smoke-filled fogs exacerbated the death rates from pulmonary conditions was becoming increasingly plain. Cold and fog were thought to be relatively harmless in rural areas, but not when married together with the urban smoke cloud. Rollo Russell reported that in a single fortnight, from 24 January to 7 February 1880, London's death rate increased from 27.1 to 48.1 per thousand due to the combined effects of smoky fog and cold. There were 2,994 deaths above the average for the time of year, while he estimated that at least another 30,000 people were made ill because of London's dense fog. The majority of the victims were 'in weakened constitutions', but by no means all of them.[250] Contemporaries observed the same phenomenon at Manchester, with John Graham noting the marked way in which 'deaths per week ... jump up during the week or two which follows a fog'.[251] The *Lancet* report concluded that it was thick, sulphurous fogs, in combination with low temperatures, that saw the adult death-rate from respiratory diseases escalate sharply upwards. Although the report did find that this trend was less pronounced in the case of children.[252] However, if children were perhaps less vulnerable to smoky fogs than the elderly, bronchitis and pneumonia were still the main killers of those under five years of age, while for infants under three months wasting diseases and diarrhoea were the principal causes of death.[253] In the decade 1891–1900 the death rate for children under the age of one was still rising, averaging 181.2 per thousand in England and Wales: and it was far worse in the 'Herodian districts' of factory towns.[254] Bronchitis was most prevalent during the winter months, and mothers caring for sick children were strongly encouraged to keep the patient warm.[255] The nursing of a sick child was one of the few occasions on which fires were lighted in working-class bedrooms. However, the remedy was part of the problem as extra coal fires in winter only served to make air pollution worse, by incrementally adding to the volume of the 'smoke nuisance'.

The causes of respiratory disease are manifold and difficult to pin down with any degree of certainty. Other variables that need to be considered in addition to smoke pollution include freezing, foggy weather, damp housing, occupational exposure to harmful substances, tobacco smoking, and the influence of sensitising agents such as moulds and bacteria.[256] Nonetheless, John Leigh, Manchester's first Medical Officer of Health, was in no doubt that coal smoke was extremely detrimental to health:

> In a letter addressed to the Registrar-General of England and Wales, in 1866, I endeavoured to show that the excessive mortality of Manchester was largely due to its vitiated atmosphere, and that the circumstances productive of such an atmosphere were mainly instrumental in causing the excessive sickness and mortality which unhappily characterise large manufacturing communities generally ... It is no answer to say that many people living in large smoky towns keep healthy and attain to a considerable age. There are strong and healthy people with great powers of resistance who form exceptions in every case, but the normal condition of old people or of those approaching old age in the working classes of Manchester is bronchitic in a greater or less degree, and men, women, and children of delicate organisation, or of scrofulous diathesis, or in whom the general conditions of health have been reduced by other causes, are very much disposed to take on phthisical or bronchial disease under the irritating effects of a Manchester atmosphere.[257]

Leigh's counterpart in neighbouring Salford, John Tatham, actively supported this point of view. In his Annual Report of 1881 Tatham argued, 'We may fairly assume, *in limine*, that a smoky atmosphere, such as we have here in Salford, is unquestionably prejudicial to health, for if medical opinion may be said to be unanimous at all, it certainly is so on this point.'[258]

Another way in which smoke pollution contributed to the growing number of deaths from respiratory diseases was through facilitating the transmission of pulmonary tuberculosis. Pulmonary tuberculosis, also known as phthisis or consumption, is a lingering, debilitating illness that is largely spread by airborne droplet infection. Although tuberculosis is not usually a highly infectious disease, people constantly exposed to the coughs and sneezes of the sick are likely to contract the disease eventually. As Figure 2 shows, mortality from tuberculosis was falling slowly in the latter years of the nineteenth century, but it still remained a major urban killer. Manchester's Arthur Ransome, one of the pioneers of preventive health measures aimed at lessening tuberculosis fatalities, repeatedly stressed that the close atmospheres of overcrowded, ill-ventilated working-class homes provided the ideal environment in which the disease could flourish.[259] In order to reduce mortality from tuberculosis Ransome advised that windows and doors should be regularly left open in order to allow a generous influx of fresh air. The 'smoke nuisance', however, proved to be an effective deterrent to adequate ventilation, even in summer, as Ransome explained in 1882:

> I wish especially to advert to the hindrance to the efficient ventilation of dwelling-houses in towns owing to the presence of smoke in the air, and to the baleful influence upon health thus produced. It is truly no marvel that our good Lancashire housewives should close their doors and windows to the air that would now pour its heavy load of soot and noxious vapour into the houses if it were admitted. After her periodical cleaning of the houseplace and bedrooms, it must be heartbreaking to a cleanly woman to see all her labour come to naught owing to the deposition of soot from the outer air; and it is hardly surprising that she will not only shut the outer windows and close every crevice between the sashes, but she will also probably stuff up the chimney with some handy plug for fear of the down-draught bringing with it further cause of offence. In itself, this practice might, perhaps, be regarded as a minor matter ... but, in reality, when its effect upon health is considered, and when we remember that its influence is felt in many thousand cottage homes, we shall see that it is of more real importance than all the other grievances put together. For it means, in reality, the encouragement, and, in many instances, the production of consumption and other forms of chest disorder.[260]

In the winter months, when warmth was a primary consideration for the poor, doors, windows, and crevices were stopped-up equally tightly to ward off cold and draughts, as well as soot. Whatever the season, the large amount of time contemporaries spent confined indoors in poorly ventilated rooms unquestionably helped to spread the 'White Plague', tuberculosis. Furthermore, if clean air for ventilation was in short supply, the absence of natural light in urban-industrial areas was also becoming a cause for concern.

By the 1880s an Italian proverb, 'Where the sun cannot enter, the doctor does', was frequently cited in Britain.[261] At the same time, Glasgow's Medical

Officer of Health James Russell attested to the value of sunlight as a 'natural disinfectant' thus:

> It has been proved by experiment that bacteria may be absolutely killed by sunlight, that sunlight is a universal disinfectant; that this disinfecting influence is exactly proportioned to the duration and intensity of sunlight, most powerful in direct, but still distinct in diffused or indirect sunlight.[262]

After it was shown by Robert Koch and others that sunlight kills the tubercle bacillus, the task of clearing Britain's skies assumed an added importance in medical circles.[263] At the sanitary congress held at Manchester in 1902 Sir James Crichton Browne's paper on *Light and Sanitation* cautioned,

> Light is a sanitary agent of the first order, and it behoves all good sanitarians to spread the light, to conserve the light, and to protect it from pollution ... over every town now, great and small, hangs a canopy of soot and noxious vapours, more or less dense, veiling the sun and cutting off much of his benign influence ... 'Let there be light' was the first creative fiat, and it had to be repeated today.[264]

The blanket of smoke that covered Manchester was also linked to the development of another serious disease – rickets. Most common in urban areas, rickets was, and is, a disease of childhood, brought about by the combination of a dietary deficiency of vitamin D and sunlight deprivation.[265] In 1889 a *British Medical Journal* investigation revealed that the incidence of rickets was greatest in large industrial towns and mining areas, and that concentrations of the disease were highest in gloomy town centres.[266] At Manchester in 1882, smoke pollution was blamed for the 'preponderance of strumous, ricketty children among the working class of the city', but it is only now that the full extent of the harm caused by the disease has begun to be analysed in detail.[267] Anne Hardy has shown that the ravages of the disease played a leading role in determining deaths from measles and whooping cough, and suggests that rickets may also have made children more susceptible to diarrhoeal, respiratory, and tubercular infections.[268] Moreover, rickets could have far-reaching effects for those that survived to adulthood, with, for example, childbirth being made more difficult for some women. In the early years of the twentieth century J.S. Taylor, Assistant Medical Officer of Health for Manchester, pointed out that as the disease left many female sufferers with a contracted pelvis, 'Gynaecologists dread the difficult labour of the woman deformed by rickets'.[269] It was believed that the death rates from puerperal or 'childbed' fever would be much reduced when smoke was abated. Contemporary observers also noted that the same factors that influenced the incidence of respiratory diseases and rickets in large towns and cities could also have serious psychological effects on their inhabitants.

Robert Roberts drew attention to the housewives who had become energetically obsessed with cleanliness in the face of an unrelenting barrage of dirt and smuts. Dark clouds of smoke, however, were commonly thought to have a depressing and devitalising effect upon the human organism. In autumn and winter, the damp and dismal British climate – considered disheartening enough in itself by many Victorians – was worsened by 'the suicidal, peasoup air', as the *Times* lamented in 1855:

> All those who have experienced the depressing effects of a November day, and have seen the atmosphere without a moment's warning put on the changeable complexion of a very bad bruise, and then resolve itself into a dull, leaden, hopeless hue, for the rest of the day, can readily understand the fixed belief of the Parisian that in that month Cockneys give themselves up to suicide, and leap in a constant stream from London-bridge. Indeed, a countryman from the breezy South Downs or from any country village where the air 'recommends itself nimbly to the senses' may well feel his heart sink within him as he looks up in vain for the blue sky, and sees nothing but that solemn gray canopy of vapour which sits like an incubus on the whole town.[270]

At Manchester the sanitary reformer Dr A. Emrys-Jones expressed similar sentiments. 'A smoky atmosphere', he commented in 1882, had 'a most depressing influence on the mind. Melancholia is directly caused by its influence, and this often leads to further mental derangements.'[271] Scientific research conducted by the psychologist Dr J.E. Wallace Wallin at Pittsburgh, reputedly the smokiest city in the United States, found that the mental efficiency of its inhabitants was impaired by atmospheric pollution. Newcomers to Pittsburgh from brighter, relatively smoke-free environments allegedly experienced 'a distinct disinclination to work, or a sort of chronic ennui'. Wallace Wallin reported a drop in both the quantity and quality of work done at Pittsburgh, including his own. He concluded that it was 'probable that irritating, acrid soot particles and poisonous smoke compounds may become factors in causing premature decay, untimely death, exaggerated fatigue, frequent sickness, instability of attention, malcontent, irritability, lessened self-control and possibly psychic disequilibrium'.[272] Wallace Wallin utilised evidence gathered at Manchester in his research, and his findings were in turn broadcast by anti-smoke campaigners in the city.[273] The Victorians, then, were aware that atmospheric pollution was not only polluting the air that people breathed, but also blocking out the health sustaining and stimulating influence of natural light. Both directly and indirectly, the smoke and gloom of the factory towns seriously affected the physical, and possibly the mental, well-being of their citizens. It seems reasonable to assume that no one in Manchester was left completely untouched by the malign influence of urban-industrial atmospheric pollution.

Accommodating smoke

Of all the factors that shaped nineteenth-century industrial cities and nineteenth-century urban life, few have been as influential – and as undervalued – as the 'smoke nuisance'. Manchester's burgeoning smoke cloud, in combination with its damp climate, prevailing south-west winds, and flat physical setting, affected every aspect of urban life, from the location of industry and middle-class housing to the drab colours of home decor and people's apparel. By the 1870s the region was so polluted that the smoke was palpable to the senses even in the countryside surrounding the industrial towns of south-east Lancashire. The urban smoke cloud, denser in winter than in summer, seriously damaged, among other things, vegetation, buildings, and the people's health. The effects on health, however, could not be proven beyond all doubt as chronic respiratory diseases have a complex aetiology. Similarly, smoke-filled fogs, which raised death rates significantly in urban-industrial districts, did not have the same dramatic impact on contemporaries as an outbreak of fever. Fogs tended to kill people who were already sick or vulnerable, whose deaths were not entirely unexpected. Town dwellers faced the acrid smoke daily, and although complaints about the polluted environment were commonplace, they slowly adapted to the harsh atmospheric conditions. Modern urban life tacitly evolved to accommodate the smoke of urban-industrial development. The city's architecture was designed with its smoke and soot in mind, while Manchester's public parks were regularly planted out with pollution-resistant species of vegetation. The city's inhabitants mitigated the effects of air pollution by spending far more time indoors, cultivating a taste for home comforts which substituted for the want of natural light and beauty out of doors.

The social habits of the people shaped, and were shaped by, the rhythms of the urban smoke cycle. Washing and cleaning activities took up more and more time in 'respectable' households, and windows and doors were kept firmly closed against dirt and smuts all the year round. With the everyday habits and rituals of domestic life being centred on the hearth, an insular, indoor lifestyle only served to exacerbate the 'smoke nuisance'. However, if much of what people discovered about their local environment came through their everyday experiences of the world, why, then, was there so little organised opposition to coal smoke? Although countless thousands unquestionably suffered from impaired breathing, smarting eyes, and smoke-begrimed skin and clothing, how people experience their environment is only one part of the story. If, as Berleant insists, environment is a 'seamless unity of organism, perception and place' the answer must also lie in the meanings that the urban-industrial culture of Victorian Manchester attributed to the city's smoke cloud, the central subject of part two of this study.[274]

Part Two

Stories about Smoke

The inhabitants of Victorian Manchester, one of the smokiest cities in the world, were constantly aware of the heavily polluted atmosphere that surrounded them. Smoke was readily perceived by four of the five senses, and, as we have seen, unpleasant sensory experiences of air pollution were part and parcel of everyday life for city dwellers. Yet, despite the tangible nature of the problem, no popular mass movement against smoke developed in the city during the nineteenth century. How can we begin to account for this seeming indifference towards such an offensive phenomenon as smoke pollution? One of the keys to explaining the lack of widespread support for smoke abatement in the city is to recognise that sensory perceptions of the dark volumes of sulphurous smoke were not simply objective responses to unpleasant stimuli.[1] There was (and is) no pure sensation as such – no disinterested 'naked eye'; only engaged perspectives that were formed from active experience of a blackened cityscape which was at one and the same time a physical reality and a shadowy *milieu* suffused with cultural values and beliefs.[2] At bottom, sensory perceptions of air pollution can not be disentangled from the particular urban-industrial context in which Manchester's citizens lived out their lives.

Although it was both harmful and onerous, contemporaries did not view smoke pollution solely as a social evil. The interpretation that townspeople placed on perceptions of smoke was in no way fixed, but depended on the circumstances in which air pollution came to the public's attention. Through the use of symbols and language cultural groups in the city were able to bestow a wide array of different meanings to the same physical object and environmental conditions.[3] In this part of the book I examine the dominant images and narratives that gave meaning to and created common understandings of a concrete environmental pollution issue in Victorian Manchester, the 'smoke nuisance'. Urban environmental degradation was rationalised and naturalised by the stories contemporaries told about air pollution. And, as William Cronon has recently argued, 'to recover the narratives people tell themselves about the meanings of their lives is to learn a great deal about their past actions and about the way they *understand* those actions. Stripped of the story, we lose track of understanding itself.[4]

Environmental discourse about smoke in Manchester was a bewildering stream of contested and contradictory claims and concerns. By analysing how a variety of actors framed the phenomenon and investigating the context in which stories about the city's smoke unfolded, we can enrich our insights into how people defined, thought, and made choices about the local environmental conditions in which they lived. Thus far Victorian urban dwellers have been portrayed mainly as being uninterested in environmental issues. However, as I shall show, the citizens of nineteenth-century Manchester were much more than apathetic spectators where smoke pollution was concerned. To bring the main story lines about smoke pollution into sharper relief, I draw on a diverse range of texts, from newspaper stories, novels, and working class autobiographies to postcards, poems, and popular songs. I focus first on a 'wealth and well-being' story line, assembling the components of a narrative that consistently emphasised the 'inevitable' correlation between smoke, economic prosperity, and a felicitous lifestyle. I then knit together the threads of a narrative that accentuated 'waste and inefficiency', constantly stressing the dangers to the health of the urban workforce, the damage to the natural and built environment, and the uneconomic and wilful misuse of Britain's finite natural resources. Finally, I suggest reasons why the concept of smoke control did not readily capture the public's imagination.

Wealth and well-being

The first of the story lines that dominated the public's understanding of the production of smoke in Victorian Manchester contended that a factory chimney and, for that matter, a domestic chimney belching out black smoke symbolised the creation of wealth and personal well-being. Most of Manchester's manufacturers, its magistrates and councillors, members of its trade associations and chamber of commerce (with two or more of these positions of authority often held by one and the same person), and its substantial workforce seemed to subscribe wholeheartedly to this narrative. The smoke was represented as a necessary by-product of industry: 'the inevitable and innocuous accompaniment of the meritorious act of manufacturing'.[5] The production of smoke warranted no apologies from most industrialists, who pointed to their smoking chimneys as a barometer of economic success and social progress. Over the course of the nineteenth century a positive, utilitarian image of Manchester's blackened physical environment had evolved, drawing on cultural values and beliefs that reflected its citizens' definition of themselves as an urban industrial workforce. For the steam-powered mills had brought material wealth for many of the city's inhabitants as well as environmental problems. Here I will map out the main features of a narrative that portrayed the 'smoke nuisance' in a beneficial light.

'The golden breath of life'

If there was one overriding image that predominated in the everyday lives of the townspeople of Victorian Britain's northern factory towns, it was the smoking industrial chimney. This image was to become indelibly stamped on the public imagination as indicative of flourishing trade and economic prosperity. At Manchester, the city's booming industries, especially cotton, provided numerous job opportunities and produced rising living standards for an ever-increasing number of its working classes, particularly after mid-century. The importance of coal and the cotton textile industry for growth and prosperity in Manchester was widely recognised in the lyrics of the popular songs of the period. In the 1840s and 1850s, for example, the comic song *Manchester's Improving Daily*, composed by Richard Baines, became a great favourite with the city's working inhabitants. The first verse went as follows:

> In Manchester, this famous town,
> What great improvements have been made, sirs;
> In fifty years 'tis mighty grown,
> All owing to success in trade, sirs;
> For we see what mighty buildings rising,
> To all beholders how surprising;
> The plough and harrow are now forgot, sirs,
> 'Tis coals and cotton boil the pot, sirs.
> Sing Ned, sing Joe, and Frank so gaily,
> Manchester's improving daily.[6]

While coals and cotton provided the workers' daily bread, this act was not usually accomplished without the 'inevitable' production of large volumes of black smoke from the city's industrial chimneys. This view of coal and smoke was widespread and is reproduced in the following verse about Glasgow:

> There's coal underground,
> There's coal in the air,
> There's coal in folk's faces,
> There's coal – everywhere;
> But – there's money in Glasgow![7]

In an 'age of smoke', popular poems and songs generated associations that helped to naturalise and rationalise the relationship between wealth and air pollution in the minds of the urban masses.

The potency of this 'natural' relationship was reinforced through periodic slumps in the cotton trade, most notably the trade depression of 1837–43, the Cotton Famine of the early 1860s, and the cyclical pattern of slumps known as the 'Great Depression' of the 1870s to 1890s. Manchester's workforce regularly experienced the hardships and misery of lay-offs, short time and wage cuts. As John Walton has observed, 'The cotton industry's growth was neither linear nor uncomplicated.' [8] During the latter decades of the century optimism about the future began to wane in the face of a growing industrial challenge from European and American competitors for a share of the international cotton market.[9] In an uncertain world the image of the smoking factory chimney became a reassuring symbol of flourishing trade and 'good times' for most contemporaries. Travelling around the cotton towns of Lancashire during the very lean year of 1842 and finding the factories of Bolton hard at work, William Cooke Taylor of Trinity College, Dublin, exclaimed,

> Thank God, smoke is rising from the lofty chimneys of most of them! for I have not travelled thus far without learning, by many a painful illustration, that the absence of smoke from the factory-chimney indicates the quenching of the fire on many a domestic hearth, want of employment to many a willing labourer, and want of bread to many an honest family.[10]

The image of thousands of smokeless chimneys, as envisioned by the anti-smoke campaigners of the period, was almost certain to cause alarm and anxiety among Lancashire's working classes. Concerns about the absence of smoke in the industrial city found expression in popular culture's representations of the dilemma. During the Cotton Famine, a cyclical slump exacerbated by the American Civil War, a poem entitled *The Smokeless Chimney* sold well, chiefly at Britain's railway stations, in aid of the Relief Fund for Lancashire's unemployed textile workers. Written in 1862 by Mrs E. J. Bellasis, under the pseudonym of 'A Lancashire Lady', it mirrors Taylor's earlier personal narration of the meaning of smoke:

> Traveller on the Northern Railway!
> Look and learn, as on you speed;
> See the hundred smokeless chimneys,
> Learn their tale of cheerless need.
>
> 5   'How much prettier is this country!'
> Says the careless passer-by.
> 'Clouds of smoke we see no longer.
> What's the reason? – Tell me why.

'Better far it were, most surely,
10 Never more such clouds to see,
Bringing taint o'er nature's beauty,
With their foul obscurity.'

Thoughtless fair one! from yon chimney
Floats the golden breath of life.
15 Stop that current at your pleasure!
Stop! and starve the child, – the wife!

Ah! to them each smokeless chimney
Is a signal of despair.
They see hunger, sickness, ruin,
20 Written in that pure, bright air.

'Mother! Mother! See! 'twas truly
Said last week the mill would stop!
Mark yon chimney, – nought is going, –
There's no smoke from 'out o'th top!'

25 Weeks roll on, and still yon chimney
Gives of better times no sign;
Men by thousands cry for labour, –
Daily cry, and daily pine.

Let no more the smokeless chimneys
30 Draw from you one word of praise.
Think, oh, think! Upon the thousands
Who are moaning out their days.

Rather pray that, Peace soon bringing
Work and plenty in her train,
35 We may see these smokeless chimneys
Blackening all the land again.[11]

Although hardly classic poetry, Bellasis's paean to air pollution (here much abridged) contains many of the cultural messages that were essential for the propagation of the myth that smoke was inextricably linked to health, happiness, and prosperity. Workers wait despondently for smoke to issue from the lifeless factory chimneys; and it is smoke, the 'golden breath of life', and not clean air, that would bring the urban masses employment, comfort, and plenty.[12]

The existence of bad trade conditions was not a prerequisite for the success of the story line that smoke was inevitable and denoted economic prosperity. At mid-century, for example, a time of neither boom nor bust in the city, the journalist Angus Reach wrote in the *Morning Chronicle*, 'Purify the air of Manchester by quenching its furnaces, and you simply stop the dinners of the inhabitants. The grim machine must either go on, or hundreds of thousands must starve.' [13] At the same time Sir James Graham, the former Home Secretary, told the House of Commons that 'smoke to an immense extent was necessary to carry on the manufactures of the country'.[14] This message was repeated unremittingly by Manchester's employers, who were increasingly worried about the squeeze on profit margins and market share in the face of foreign competition. Their views, closely associating smoke with continued economic growth, were regularly reported at length in the *Manchester Guardian* and other local newspapers. Reginald le Neve Foster, an influential director of Manchester's Chamber of Commerce, countered one of the City Council's many attempts to enforce the law against smoke pollution by declaring that if it succeeded, 'they would drive away all their industries, ... and Manchester would soon become one of the "dead cities" of the world'.[15] This strand of the narrative of wealth and well-being had a wide currency and found a receptive audience beyond the working people of the Lancashire textile towns. At Sheffield William Nicholson, Chief Smoke Inspector, whose job it was to abate air pollution from its numerous steelworks, cautioned that,

> There could be permanently smokeless cities, and a smokeless country, if all furnaces and fires were put out and never relighted, but a sensible people would never demand or even desire such a senseless thing, for a smokeless country, even with its purer air, clear skies and more sunshine would be a country of universal poverty, but a country with its furnaces and fires making only the necessary smoke for the carrying on of industries, though a little smoky, would be a country of universal prosperity ... what cannot be cured must be endured, without grumbling, for it is a lesser evil than putting out the furnaces and fires.[16]

This was a narrative that was to a large extent shamelessly predicated on negative images of smokeless chimneys. It played constantly and effectively on immediately intelligible fears about what life in the industrial city would be like *without* its familiar and reassuring smoke cloud. Just as today, the issue of pollution control was often viewed in simplistic terms, with the manufacturers presenting a stark choice between smoky prosperity or economic stagnation if environmental safeguards were proposed. From the 'low' culture of the ordinary working people to the 'highbrow' discussions of politicians and businessmen, the powerfully persuasive message was that the presence of industrial smoke actually sustained a thriving industrial community.

   On a more modest scale, a generous amount of smoke seen freely issuing from any one of Manchester's many thousands of domestic chimneys signified a working family's continued good fortune. Mrs A. Romley Wright, who taught domestic economy classes in the city, illustrates the symbolic power of the smoke emitted from 'the popular British institution' of the open coal fire: 'The kitchen fireplace is filled with coal – large pieces, of course, for roasting. A volume of smoke rushes up the chimney, and the admiring neighbours may ejaculate "Oh, *what* a dinner Mr so and so must be having."" [17] An extravagantly smoking chimney pot visibly demonstrated to onlookers that a family was doing well economically, and might even have enhanced their social status in the community. The smokeless fuel coke, although relatively inexpensive, was unpopular among the city's inhabitants. Coke did not make a good blaze in the hearth and was widely perceived as 'a fuel of poverty'.[18] As was the case with industrial chimneys, smoke spouting from domestic chimney pots was a symbol commonly employed by contemporaries to indicate 'good times'. The other side of the coin – the cold, fireless grate – was an image that was used by novelists of the period, from national figures such as Charlotte Brontë, Elizabeth Gaskell, and Charles Dickens to local working-class dialect authors such as Edwin Waugh and Ben Brierley, to denote want and poverty. And when, for example, Waugh wrote of 'fireless hearths, an' cupboards bare' in the song *Hard Weather* (penned during the acute recession of 1878–79), he would without question have sent a pang of anxious recognition through many of those who heard or sang it.[19] There were, however, different ways of seeing smoke that drew upon powerful concepts other than the growth of trade and economic prosperity.

'Fireside bliss'

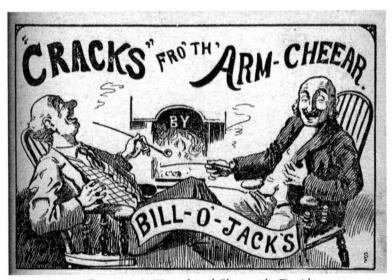

Illustration 5. Warmth and Cheer at the Fireside.

Source: William Baron's *Bill-o'-Jack's Lancashire Monthly*, May 1911.

Domestic life in Victorian Britain revolved around the fireplace, especially during the cold, damp and dreary winter months. In freezing weather 'toasting' oneself before the warm, radiant open fire was an activity commonly indulged in and enjoyed by all social classes. However, a blazing coal fire imparted much more than an agreeable degree of heat. Robert Angus Smith, perhaps nineteenth-century Britain's foremost scientific authority on air pollution, recognised that domestic fires provided a great deal besides physical warmth for contemporaries when he wrote, 'coal smoke … is associated in our minds with pleasant ideas of comfort rather than of nuisance'.[20] The domestic hearth was associated with the notion of human warmth, signifying love, friendliness, and a sympathetic, comfortable environment. Moreover, the mantelpiece above the hearthstone was also important in this respect. It was customarily used to display treasured possessions that bound the family together.[21] Besides ornaments, mementoes, and family photographs, these prized belongings sometimes included framed mottoes which promoted, often in mawkish language, the idea that love and security were to be found at the fireside. 'Warmth and cheer attend you here'; 'A world of care shut out, A world of love shut in'; and 'Where friends meet, hearts warm' are but three examples among a whole host of similar sayings.[22]

There are innumerable popular images, both literary and visual, extolling the pleasures of hearth and home that date from the Victorian period. Charles Dickens, who was an active opponent of the 'smoke nuisance', nonetheless penned many scenes of snug domesticity that eulogised the 'cheerful' open coal fire. In *The Cricket on the Hearth*, for instance, the cold, wet, and exhausted John Peerybingle returns home after a hard day at work to find a 'jolly blaze' in the hearth of his family's 'snug small home'. The boiling kettle on the hob, and the cricket, cheerfully add their welcoming 'fireside song of comfort', and the weary worker is soon ensconced in 'domestic Heaven'.[23] The dialect author Ben Brierley often depicted working people gathering in a crescent around the hearth – 'fro' hob to hob' – to form a 'gradely Lancashire "fender"'. On these occasions several generations of a family and friends would crowd together in front of a bright fire to sing songs, crack jokes, and tell ghost stories late into the evening.[24] The genial glow of the fireside could also be enjoyed outside the confines of the 'respectable' working-class home. Numerous dialect tales are set in the cosy, companionable surroundings of the inglenook or chimney corner of the local public house.[25]

The open coal fire was represented almost as a living thing that responded to and conveyed every human mood. The crackle and roar of the leaping flames or the soft purring of the glowing embers on the hearthstone were vibrant with meaning for contemporaries. The blazing fire was 'a grand depository of images' for those inclined towards solitary reflection, as the musings of William Trevor reveal:

> I am in one of those moods to-night when the fire, as it glows and splutters and flickers, has a great fascination for me. It is just gloaming, too early to light the gas, and too dark to read; so I draw my chair towards the fire, place my feet cosily on the fender, and, as I fold my hands leisurely, I fling the reins on the neck of my imagination. I people every glowing ember with some long-forgotten face, whilst every tongue of flame talks to me of happy boyhood as it reflects the rosy hues which tinged the scenes of life's young fancies.[26]

In 1867 L.M. Budgen, in the book *Live Coals; or, Faces From the Fire*, actually attempted to interpret the 'figures' or 'types' which became visible in the flames for the general public. More importantly, however, Budgen explicitly brought to the fore many of the positive, deep-seated associations between well being and the smoky domestic coal fire:

> But *the* fire! – the dear familiar fire, that lights up our hearth, and faces round it! What is *that* to you, and me, and all of us, who have a fire (God help, and we too, those who have none!) to call their own? Nobody, – nobody, at least, except dwellers between the tropics and patrons of Arnott stoves, – need be told that the fire which looks out at us so cheerily from the bars is the companion of the solitary, the comforter of the sad, the enlivener of the dull, the magnet of social attraction, the pivot and cherisher of tender recollections; in a word, the sun (when the summer sun is wanting) of every domestic system: hence its very *life*; not forgetting its vulgar but particularly *vital* uses, as roaster of the joint, and boiler of the kettle.[27]

The coal fire was 'a live thing in a dead room', without which home life would become dull and miserable.[28] The popular culture of the day, then, often depicts a family and friends seated around a roaring fire that *on its own* created a most congenial atmosphere. A verse from the Lancashire dialect writer Edwin Waugh's short poem *Toddlin' Whoam* encapsulates the powerful attractions of the domestic hearth:

> Toddlin' whoam, for th' fireside bliss,
> Toddlin' whoam, for th' childer's kiss;
> God bless yon bit o' curlin' smooke;
> God bless yon cosy chimbley nook!
> I'm fain to be toddlin' whoam.[29]

Although such representations of hearth and home were often overly romanticised, it would be extremely pessimistic to suppose that such pleasant activities were not experienced by most working-class families, at least from time to time.

According to this strand of the story line, Victorian Britain's homespun pleasures would be very much diminished without the 'cheerful' domestic coal fire as an animated backdrop to everyday social activity. As Professor William Bone, who established the Department of Fuel Technology at the Royal College of Science in 1912, told the assembled ranks of the National Health Society,

> ... we must all confess that there is something indefinably pleasing about the ever-changing aspect of a coal fire, with its dancing flames and genial warmth, which not only appeals to our 'fire-worshipping' instincts but is also well suited to the character of our British winters with their sombre skies, searching fogs, and rapid changes in temperature ... a fire is a welcome companion ... And to all of us the sitting round it is one of the most cherished features of our home-life. In abolishing it we might save coal, but we should lose England.[30]

The sights and sounds associated with the traditional open fireplace emphatically denoted warmth in every sense, and it had become an integral part of the national culture. Furthermore, even the disagreeable, sulphurous odour of emissions from this 'popular British institution' could be portrayed as having positive connotations for the Victorian city-dweller.

Disinfecting the urban atmosphere

The streets of nineteenth-century Manchester reeked from a chaotic combination of pungent odours, originating from both domestic and industrial sources. From the stinking human and animal excrement and decomposing refuse that littered the streets to the acrid emissions of smoke and noxious vapours from the city's diverse industries, strong smells were an integral part of the urban-industrial environment. In a hierarchy of the senses, however, smell, along with taste and touch, was ranked well below vision and hearing by philosophers such as Immanuel Kant.[31] The visual and aural faculties guide human actions on the basis of 'civilised' intellectual reasoning, while 'smelling and sniffing are associated with animal behaviour'.[32] But the polluted surroundings of many of Manchester's poorest inhabitants were far from civilised, even towards the end of the nineteenth century, and the sense of smell was heavily relied upon as the first line of defence against numerous harmful substances in the urban atmosphere.[33] Alain Corbin has revealed 'a collective hypersensitivity to odours of all sorts' in eighteenth and nineteenth century France, a position that was closely paralleled in the crowded districts of towns in Victorian Britain.[34] This preoccupation with odours is easy to understand when one takes into account that the pre-Pasteurian miasmatic theory of disease, which was dominant until the 1880s, held that foul smelling air from putrefying matter actually caused sickness, disease, and death.[35] Long-lived narratives and popular attitudes concerning the odorous 'smoke nuisance' developed in an environment charged with fear, as city air was believed to be thick with dangerous miasmas. To paraphrase Roy Porter, the stenches of the new factory towns not only filled the noses of their inhabitants; they also filled their minds.[36]

In the 1840s the notion of miasma pervaded important national reports on the 'condition of England' question.[37] Edwin Chadwick and other like-minded Victorian sanitarians represented the putrid smells that rose from sewer-rivers, cesspools and drains, over-crowded graveyards, slaughter-houses, and other decomposing organic matter as the major sources of the miasmas which were thought to cause so much illness and mortality.[38] Moreover, scores of Victorian fiction writers frequently connected the disgusting smells of the polluted urban environment with fever and sickness. In *Bleak House* Charles Dickens described in menacing terms the 'nauseous air' and 'pestilential gas' in which the unhealthy slum dwellers of Tom-All-Alone's lived. This contaminated air, when swept along by the prevailing winds, he warned, had the potential to harm 'every order of society'.[39] Dickens' frightening scenario exemplifies pressing contemporary anxieties about infection from a tainted urban atmosphere. To rid towns and cities of malodorous filth and ordure before it could breed disease and to secure adequate supplies of clean water for drinking and flushing sanitation systems were the sanitary reformers' main priorities, not the abatement of the 'smoke nuisance'. For, as Elizabeth Gaskell noted in depicting the fetid atmosphere of a Manchester slum in *Mary Barton*, 'The very smoke seemed purifying and healthy in the thick clammy air.' [40] If smell was placed at the bottom of a hierarchy of the senses, coal smoke likewise rated a low ranking in the hierarchy of noisome odours.

Manchester's smoke, which offended the senses and inhibited the breathing of its citizens, was, paradoxically, widely believed to be beneficial to health. This component of the narrative maintained that the carbonaceous particles and the sulphurous and sulphuric acids existing in the city's polluted atmosphere protected against miasmas and disease. The deodorising properties of soot and 'smuts' were highlighted by manufacturers as an important weapon in the battle against excremental and other foul odours. Similarly, the fumes from burning sulphur had long been used to fumigate and disinfect the air of contaminated places.[41] Thus, sulphurous and sulphuric acids, although considered harmful to people's health when present in the urban atmosphere in high concentrations, could also be depicted as powerful agents of purification. In 1857, for example, the alum manufacturer Peter Spence's paper on coal smoke, delivered to the Manchester Literary and Philosophical Society, argued that soot,

... the solid carbon, against which all the cry is raised, is guiltless of any deleterious effect on human health, [it] is one of the most anti-putrescent bodies, and while floating in the atmosphere does all that it can to arrest and destroy noxious and miasmatic vapours ... [and] sulphurous acid gas ... [is] in itself one of the best correctives of miasma, yet certainly a most insalubrious atmosphere to breathe constantly.[42]

A contributor to the journal of the Society of Arts, the prestigious body which had organised the impressive display of industrial achievements that was the Great Exhibtion of 1851, reiterated this concept during a discussion on air pollution:

He would put in a plea on behalf of soot. In his opinion, the carbon which was deposited on the roofs of the houses in the form of soot, and afterwards washed away by rain, and conveyed into the sewers, acted, in a great degree, as a deodoriser of the sewage matters, and in that way was beneficial in a sanitary point of view.[43]

Towards the end of the century the Earl of Wemyss informed the House of Lords that 'what rendered the air of London, with its 4,000 miles of sewage, less deleterious than it might be expected to be was the great quantity of carbon with which it was impregnated'.[44] And as the ashes from domestic coal fires were used every day by Britain's city-dwellers to deodorise the stench from countless dry-pan privies, it is likely that this idea seemed valid to many contemporaries.[45] Soot was also utilised to good effect by the Victorians as both a fertiliser and an insecticide;[46] not forgetting that it was used as toothpaste and ingested by the working classes as a home remedy for stomachache.[47] Moreover, the acidic fumes which abounded in the smoky air of the northern factory towns were commonly thought to protect against illness.

While in no way being the most important strand of the story line of wealth and well being, the notion that smoke was beneficial to health was nevertheless made much of. In 1842 the *Manchester Guardian* reported the view of many of the city's businessmen: 'if we had not smoke we should have typhus'.[48] In 1859 Robert Angus Smith, at this time earning his living as a consultant chemist in Manchester, reinforced this message when he wrote:

There can be little doubt that ... the sulphurous acid of the coals ... acts as a disinfectant of the putrid matter in towns, and anyone with an attentive smell passing through the streets of this country and the cleanest towns of the continent must feel how vastly superior our atmosphere here is in respect of putrid matter capable of affecting that sense: whilst there is a great charge against the sulphur, we must not omit to speak so far in its favour ... It seems to me that the statistics of epidemics in large manufacturing towns fully bear out this belief, although a few cases present difficulties ... [49]

Michael Sigsworth and Michael Worboys have recently shown that the working classes actively supported the claim that acidic smoke afforded protection from disease. During an outbreak of cholera at Leeds in 1849, many local people were quick to take advantage of the town's 'free atmospheric disinfectant'.[50] Some townspeople ignored calls from the authorities to clean up the dirt and filth outside their homes, choosing instead to stand immersed in the smoke of a nearby alkali factory for days on end.[51] The acidic fumes which issued from the chimneys of chemical works were far more caustic than ordinary coal smoke. Nonetheless, the actions of the local inhabitants are still a clear indication that this concept was a significant part of the public's understanding of air pollution issues.

These popular beliefs about the purifying power of coal smoke also survived the mid-Victorian revelation that microbes, not miasmas, caused sickness and death. By the 1880s the new bacteriology or germ theory was slowly beginning to displace the previously accepted notion that the foul odours of putrefaction were the causative agents of epidemic disease.[52] However, old ideas about disease transmission were not simply abandoned overnight by the scientific and medical communities or, for that matter, by the general public. At Manchester the physician John Thresh's report into the appalling mortality of No.1 sanitary district, Ancoats, published in 1889, is notable for the way in which it constantly stresses the malign influence of decomposing organic matter.[53] The city's Medical Officer of Health, John Leigh, does not mention the germ theory of disease in his annual reports until 1887, and even then he did not exclude the possibility that 'putrescent odours' could still be a cause of disease.[54] As Nancy Tomes has pointed out, 'new information about microorganisms was understood and acted upon in the framework of older ideas and behaviours'.[55] Disease and decomposition were still strongly linked, with stinking excrement and rotting refuse often being depicted by modernist sanitary reformers as fertile breeding grounds for dangerous microbes.[56] Stench continued to communicate the presence of disease.[57] Well into the twentieth century both the miasmatic and germ theories of disease could still be found interwoven in popular sanitary advice literature, as many hygiene writers thought it best to include all possible threats to the health of a household.

However, whichever theory of disease was applied by contemporaries, the dual capacity of soot-laden and sulphurous coal smoke to deodorise and disinfect meant that it could be portrayed as either the neutraliser of foul odours or the destroyer of germs, thereby satisfying both bodies of opinion. In 1892, for example, an editorial article in the *Lancet* pointed to smoke's effectiveness against both modes of disease transmission. Commenting on the declaration by the president of the Institute of Civil Engineers that 'the combustion of coal is most beneficial to the health of the inhabitants of London', the *Lancet* stated:

> As many as 350 tons of sulphur are thrown into the air in one winter's day, and the enormous quantity of sulphurous acid generated from it deodorises and disinfects the air, destroying disagreeable smells emanating from refuse heaps and sewers and killing the disease germs which find their way into the atmosphere. There may be a good deal of truth in this view, but there is undoubtedly another side to the question ... [of the] sulphurous acid in London fogs, for although it may be beneficial to the London householders by destroying microbes, it certainly frequently does them harm by attacking their lungs and bringing on bronchitis and asthma which sometimes prove rapidly fatal, to say nothing of the minor discomforts of a disagreeable taste, filthy smell, stuffed nose, husky throat, smarting eyes, and headache. We think that, healthy though the London fogs may be, the discomforts they cause are so great that Londoners would be really better without them, and that less disagreeable and equally efficient means might be found to clear the air of microbes ... [58]

The *Lancet* argued that in this particular case the remedy was perhaps as harmful to health as the rank smells and deadly germs themselves. Nonetheless, given that filthy streets and the slow removal of human wastes and refuse remained a major pollution problem in many late-Victorian towns, it was not difficult for adroit smoke producers to go on presenting their smoke and soot as being positively beneficial to the people's health. At the same time, the poor sanitary conditions of slum districts enabled some contemporaries to argue convincingly that cities like Manchester had more urgent environmental problems to deal with than that of the 'smoke nuisance'. In 1891 the *Manchester Guardian* reported an address made by the leading businessman Reginald le Neve Foster to the local section of the Society of Chemical Industry, in which he boosted the smoky atmosphere of Clayton, adjacent to Ancoats:

> It was a remarkable fact that the Clayton district happened to be the most healthy in the city of Manchester, and that not only did they get their own smoke, but for two-thirds of the year during the prevailing winds they received a large contribution from the city of Manchester, thus proving conclusively that smoke abatement was of less vital importance to the well-being of the community than other sanitary reforms. [59]

Cleansing Manchester's Augean stables – the city's overcrowded slum districts – of 'stinking horrors' should be the main task of the local authorities, and not the eradication of relatively wholesome coal smoke. [60]

The fear of disease-carrying miasmas remained strong among the city's inhabitants, and the foul odours emanating from blocked drains, defective sewers, over-flowing privies, and the great mounds of human excrement collected at Manchester Corporation's 'nightsoil' depots at Water Street and Holt Town were the cause of great concern.[61] In 1885, for example, an anonymous resident of Ancoats complained about the 'abominable stenches' emitted from the district's Holt Town depot, which s/he characterised as 'the Death Works of our Health Department'.[62] As infectious microbes were now also understood to thrive in accumulated filth and dirt, it was becoming clear to contemporaries that the city's over-stretched dry conservancy pail system of sanitation, with its noisome storage depots, needed to be improved upon. The introduction of a water closet system was thought to be the cheapest and most effective way of quickly getting rid of wastes injurious to human health, and in 1892 an application was made to the Local Government Board for an order to undertake the conversion.[63] However, as late as 1911 less than half of Manchester's privies were water closets, with a large number of substandard midden-privies and pail closets continuing in use.[64] In the early years of the twentieth century the city's unsatisfactory dry-closet system was still producing large volumes of malodorous excrement that needed to be collected and disposed of on land. In common with many other northern industrial towns, Manchester had been slow in providing working-class homes with water closets and in constructing an efficient sewage treatment plant. Against this backdrop, a narrative that represented acidic smoke as beneficial to health could still be broadcast to good advantage at the turn of the twentieth century.[65]

Moreover, the everyday washing and cleansing practices of the Victorians must have made this notion even more compelling. Very similar substances to those that were thought to disinfect the urban atmosphere were widely used to purify domestic living spaces. From the 1840s, many of the strong chemical disinfectants and soaps used daily by 'dirt-chasing' housewives and servants in laundering clothes and scrubbing the home clean, such as carbolic acid, were produced from the distillation of coal tar.[66] What is more, the old odoriferous technique of burning sulphur liberally in sickrooms to protect the rest of the household against infection proved enduringly popular, surviving down into the early decades of the twentieth century. The 1906 edition of Mrs Beeton's *Household Management*, for example, still advocated the burning of sulphur for such purposes.[67] Margaret Horsfield has recently described this kind of fumigation taking place as late as the mid-1920s at her grandmother's home, following a case of scarlet fever:

> A public health notice warning of the infection was mounted forebodingly in the front of their house, no visitors were allowed, and great was the fear in the

household. No one but my grandmother went near the sufferer during his illness. When he recovered my grandmother went, coughing, from room to room carrying a scuttle full of hot coals sprinkled with sulphur, fumigating the entire house to prevent her other children from inhaling the deadly germs.[68]

The popular domestic manual *Cassell's Household Guide* recommended the use of 'milk of sulphur' to disinfect the clothing of the recovering patient.[69] In addition, Edwardian mothers are known to have exposed their sick children to the sulphurous fumes of the household gas meter as a home-remedy for whooping cough.[70] Less serious ailments were also treated with sulphur-based preparations. Alfred Fennings' *Everybody's Doctor* prescribed taking a teaspoon of 'flowers of sulphur' in treacle or milk for the treatment of skin diseases, while an advertisement for 'Sulpholine Lotion' placed in the *Manchester Guardian* promised to cure, among other things, eczema, psoriasis, pimples, blotches, and acne spots.[71]

With regard to smoke abatement, then, it is perhaps no surprise to find the *Westminster Review* in 1904 bemoaning the existence of 'an obstructionist party who say that coal-smoke is a disinfectant, and that, without it, infectious diseases *might* be more prevalent than they are'.[72] Deep-rooted social practices and fear of the reeking, insanitary environmental conditions intertwined with the narrative of well-being to place 'purifying' coal smoke firmly on the bottom rung of a hierarchy of offensive smells in the minds of contemporaries. The acrid smell of smoke, unlike the foul odours associated with miasmas, did not fill the Victorian city dweller with apprehension. Indeed, many people actively embraced the idea that this form of air pollution deodorised and disinfected the urban atmosphere. While by no means viewed as conducive to health by one and all, it is not difficult to understand why, in this particular context, heavy urban emissions of sulphurous black smoke could be commonly perceived to be relatively salubrious. And if these cultural beliefs meant that the odour of coal smoke was not altogether repellent to the nose, was it also possible to represent the image of Manchester's degraded, blackened cityscape as pleasing to the eye?

'Beautiful Manchester'

Some contemporaries did find beauty in the 'Rembrandtesque effects' of the smoky atmosphere.[73] In his observations on the industrial towns clustered around Manchester in 1842, William Cooke Taylor noted that their smoking factory chimneys, when not too numerous, 'produce[d] variations in the atmosphere and sky which, to me at least, have a pleasing and picturesque effect'.[74] Moreover, during the nineteenth century a number of important works of art were produced that drew inspiration from the 'atmospheric variations' caused by smoke and 'pea-souper' fogs. J.M.W. Turner, one of Britain's most original painters, was 'fascinated by steam and fog', while Claude Monet, a founder of impressionism, painted his Thames series during London's foggy winter months.[75] The urban smoke cloud and related dense, coloured fogs also found favour with some Victorian and Edwardian authors. Bram Stoker, in his famous Gothic novel *Dracula*, gave a detailed account of the 'marvellous tints' of the 'wonderful smoky beauty of a sunset over London'.[76] And Arnold Bennett, in *Anna of the Five Towns*, fulsomely praised the 'wreaths of yellow flame with canopies of tinted smoke' that illuminated a 'romantic summer night' in the Staffordshire Potteries.[77] The author of *Black's Guide to Manchester* likewise described that city's smoke in a romantic fashion when commending the view from Queen's Park back towards the city: 'there is a wide prospect of roofs, spires, and chimneys that sometimes loom through the sunlit smoke with quite Turneresque effect'.[78] No less an aesthetic authority than John Ruskin praised the 'exquisite beauty' of 'fine silvery blue' smoke when viewed issuing from the chimney of a secluded country cottage, preferably surrounded by trees. 'Now', he wrote, 'there is no motion more uniform, silent or beautiful than that of smoke'.[79] But the picturesque smoke that so captivated Ruskin's eye when seen in simple, rustic isolation held no such attraction for him under very different circumstances.

The gloomy factory towns of Lancashire and Yorkshire, with their countless tall chimneys 'foaming forth [a] perpetual plague of sulphurous darkness', were to Ruskin 'depressing and monotonous' aesthetic wastelands.[80] And Ruskin was not alone in his vociferous condemnations of the grey and dreary aspect of the manufacturing districts that were encroaching upon, and polluting, ever larger areas of Britain's countryside. When distributing prizes at Manchester's School of Art in 1885, the artist Valentine Prinsep RA observed that 'the sadness of Manchester colour was a grief to every artist who entered the city; that nothing here recalls the "joys of life"; you are reminded of nothing "but money making"'.[81] The socialist poet and designer William Morris, who came under the influence of Ruskin at Oxford, similarly attacked the ugliness of the modern centres of industry.[82] In his Utopian work *News from Nowhere*, Morris has Manchester – the spiritual home of air pollution – disappear without a trace from the face of the earth.[83] John Ruskin's aesthetic views were extremely influential in shaping nineteenth-century attitudes towards architecture and the natural environment, and very few people attempted to defend smoky Manchester by stressing its inherent beauty. Smoke, pleasing to the eye when seen ascending skywards from the solitary chimney of a country cottage, rather lost its charm for most Victorians where active chimneys could be counted in their thousands. However, with regard to Manchester's blackened physical environment, a positive, utilitarian image of the city's unsightly face had evolved: an image that drew upon, and which reflected, its inhabitants' understanding of the city as a working, industrial landscape.

If the monotonously flat character of Manchester's shrinking natural environment was unimpressive, in the first decades of the nineteenth century its expanding futuristic cityscape was the focus of global attention. Manchester's spectacular transformation from a predominantly verdant and countrified town at the end of the eighteenth century into 'the classic type of a modern manufacturing town' by the 1840s had generated an exciting new image of the place as a powerhouse of industry, progress, and wealth creation.[84] An admiration for the technological advances of the early industrial world is evident in Disraeli's description of the 'mighty region' of Manchester in *Coningsby*, where functional factory buildings have 'more windows than Italian palaces, and smoking chimneys taller than Egyptian obelisks'.[85] Manchester, the 'great city ... of smoke and toil', was represented by Disraeli to fascinated contemporaries as nothing less than the modern counterpart of ancient Athens. It is, therefore, not unreasonable to assume that ordinary working men and women must themselves have taken some small measure of pride in being a citizen of 'Cottonopolis', a distinctively modern city whose steam-powered textile industry was attracting world-wide interest. As the century wore on Manchester's fast-growing workforce no longer expected to see 'the gayest side of nature' in their gloomy, workaday environment.[86] To the despairing gaze of Romantic 'outsiders' like Ruskin, Prinsep, and Morris, the smoke-begrimed fabric of Lancashire's factory towns provided a nightmarish glimpse into the barren, polluted, and mechanised future.[87] In contrast, from the standpoint of the hardworking members of these urban-industrial communities, their treeless, coal-black surroundings could be readily perceived as emblematic of the tireless activity and Herculean energy of a region that was at the hub of the Industrial Revolution. The journalist and socialist Allen Clarke, himself a former factory worker, summed up the attitude of Bolton's inhabitants towards their drab industrialised environment thus: 'The[y] ... would laugh if told the real waste of the land began when cotton mills were planted upon it. Their eyes see nothing very hideous or sad in their town.'[88]

The novelty of Manchester's cityscape, dominated by its huge 'palaces of industry' and tall, triumphal chimneys, diminished after mid-century as other towns and cities acquired extensive factory industries of their own. Nevertheless, even in the 1910s a local guidebook could argue that the region's cotton mills were still an unmissable attraction:

> The manufactories in and about Manchester are naturally the great sights of this part of Lancashire and few things are fuller of throbbing interest than the spinning and weaving mills whose exteriors are so inartistic.[89]

The design of Victorian factories and tall chimneys, for the most part, cannot be said to have presented much for the eye to admire.[90] Indeed, to be 'as hideous as a factory chimney' had become a byword for ugliness.[91] However, what these soot-encrusted monumental structures did manage to convey was a sense of progress, strength, and power. Manchester's unsightly black face was able to command respect as 'a face full of character, showing great tenacity of purpose'.[92] According to the early twentieth century travel writer H.V. Morton, Manchester was 'a man among the cities of the earth' and its 'jet-black' countenance inspired loyalty from its citizens as well as respect.[93] The fuliginous layer of carbon that coated the city's fine public buildings and grand warehouses undoubtedly detracted from their aesthetic value. Yet to many contemporaries the dour black edifices signified enterprise and industry – the very essence of Manchester's *raison d'être*.

The Victorian's well-documented abhorrence of dirt and filth did not always extend to coal smoke, which was often portrayed as good, honest dirt and not as 'matter out of place'.[94] Indeed, that Manchester's smoking chimneys came to be widely interpreted as benign signs of progress and prosperity is also indicated by a northern expression that has survived to this day, 'Where there's muck, there's brass.' As a result, the black exterior that the hard-working city of Manchester presented to the world could be viewed uncritically by many contemporaries. In 1887, for example, a contributor to the *Manchester Guardian* wrote:

> ... as beauty is only skin deep, ugliness may possibly repose upon the same fragile foundation ... Physically, we must admit that Manchester does not make a good show, except of dirt; but it is only work-day dirt after all – the grime of a collier who has to deal with coal, the dust of a miller who has to apologise for his floury proportions.[95]

Smoke was represented as an innocuous, even beneficial, form of dirt that constituted no great threat to life and health. By the turn of the twentieth century affectionate, tongue-in-cheek images of the city's smoking factory chimneys were appearing on postcards bearing the legend 'Beautiful Manchester'. The image of a smoking industrial chimney had become as comforting psychologically as that of the domestic hearth. The production of smoke was commonly understood and hailed as an infallible sign that Manchester was a flourishing and enterprising city.

Illustration 6. 'Beautiful Manchester'

Early twentieth century postcard in the collection of Stephen Mosley.

Throughout the nineteenth century and beyond, the narrative of wealth and well-being continued to resonate with meaning, with Reuben Spencer, chairman of the board of Rylands & Sons, Manchester's largest cotton textile company, writing of the city in 1897:

> The factories are still there, the 'incense of industry' still floats in clouds above the tops of countless towering chimneys, the throngs of busy workers are more numerous than ever, and the whirr of gearing and the hum of machinery resounds in a hundred streets, whence emanate a thousand different wares for the use and benefit of the peoples of the earth.[96]

As late as 1913 *Black's Guide to Manchester* told of the city's 'thick cloud of smoke that turns to invisible gold'.[97] The story line was imparted through a great variety of texts in Victorian Manchester, all of which bracketed smoking chimneys with healthy trade conditions and stable or rising living standards. Charles Dickens sardonically captures the spirit of the age when he has Josiah Bounderby of Coketown say, 'First of all, you see our smoke. That's meat and drink to us. It's the healthiest thing in the world in all respects, and particularly for the lungs.' [98] But despite the willingness of the majority of Manchester's manufacturers, politicians, and workers to endure polluted air in the name of growth and prosperity, an active minority of influential reformers questioned the popular belief that smoke was synonymous with economic and social progress and countered with a compelling story line of their own.

Waste and inefficiency

The second of the story-lines that conferred symbolic meaning to the production of smoke reflected the values and beliefs of a largely middle-class, educated and professional elite, who, rather than viewing smoke as signifying prosperity and progress, saw the columns of sulphurous black smoke as 'barbarous' signs of waste and inefficiency. Doctors, engineers, architects, lawyers, clerics, and others from the burgeoning professional ranks, along with several of Manchester's leading merchants and manufacturers, all promoted this sceptical alternative narrative. From the 1840s on, many reformers banded together to form anti-smoke societies in the city, among which were the Manchester Association for the Prevention of Smoke (1842) and the Manchester and Salford Noxious Vapours Abatement Association (1876). The anti-pollution activists challenged entrenched cultural values and beliefs about Manchester's 'productive' smoke by holding public meetings against air pollution; by regularly inviting leading 'experts' to lecture on the subject; by testing and exhibiting smoke abatement technology; and by publishing articles and letters in newspapers, magazines, and journals.[99] Coal smoke, according to this story line, meant a failure to make profitable use of valuable and finite natural resources and a reckless waste of irrecoverable energy. Smoke meant the needless defacement and destruction of the city's buildings and green spaces, it signified an unnecessary and preventable loss of life and health, and, finally, it represented a serious threat to Manchester, Britain, and empire. The narrative that smoke was synonymous with waste was a denser, more complex response to the dilemma posed by smoke pollution and was conveyed in considerably fewer, and often less accessible, texts. However, although much of what follows is reconstructed from sources that did not enjoy an extensive popular readership, these narratives were widely disseminated in both the local and national presses.

The 'Black Smoke Tax'

Sensory experience of local environmental conditions featured prominently in the emerging story line that questioned the validity of established beliefs about 'productive' smoke pollution. Visual evidence garnered by Manchester's anti-smoke campaigners had first galvanised them into challenging the 'general allegation' that smoke abatement was 'impracticable; and ... utterly hopeless'.[100]

Common observations of Manchester's numerous tall chimneys, which conveyed its 'inevitable' emissions of coal smoke quickly and cheaply into the 'vast atmospheric ocean' overhead, had revealed to keen-eyed contemporaries that 'some chimneys emitt[ed] a much less quantity and less dense volume of smoke than others'.[101] The reformers argued that what was seen to be accomplished by a minority of Manchester's manufacturers could with a little effort be achieved by all, and that 'three-fourths, and even with care nine-tenths, of the smoke might be consumed'.[102] In a lively editorial on the topic the *Manchester Guardian*, with some reservations, gave strong support to the reformers' position:

> It is with great pleasure that we direct the attention of our readers ... to plans for the prevention of smoke, with the view of removing a nuisance and an opprobrium under which Manchester labours more heavily perhaps than any other town in her majesty's dominions ... Of course, if this depravation of the atmosphere of Manchester, by clouds of dense smoke, is a necessary consequence of the manufactures carried on in the town, the evil, however much it may be regretted, must be patiently borne; but we are satisfied – and, we believe, almost every body is satisfied – that no such consequence need follow from the existence of manufactures. Every one who has paid the least attention to the subject is well aware, that, whilst the majority of the large chimneys in the town vomit forth clouds of smoke which actually obscure the sun, others, belonging to establishments where quite as large a business is carried on, are perfectly unattended by any such nuisance, and indeed emit little more smoke than the chimney of an ordinary kitchen. Now, what can be done by one man can also be done by another who will adopt the same arrangements, and exercise the same care in working them out; and, that being the case, why should the inhabitants of Manchester continue to be annoyed, and to have their clothes and their habitations defiled by the dense clouds of smoke with which improvident and careless people will persist in polluting the atmosphere, as much to their own loss, as to the discomfort and detriment of their neighbours?[103]

The *Manchester Guardian* set the tone for debate throughout the century in condemning the negligence and irresponsibility of the majority of the city's factory owners. The newspaper's disapprobation of both smoke producers and smoke production was constantly echoed by campaigners down the years, becoming a core component of the narrative of waste and inefficiency. To the Victorian mind the squandering or misuse of material resources that could be turned to good account was nothing short of shameful. As William Garden Blaikie, successful popular author and Professor of Theology at New College, Edinburgh, put it, 'All of us have an instinctive dislike of waste, and an instinctive satisfaction in the recovery of lost or waste material, of whatever kind, and its application to useful purposes.'[104] Indeed, thrift and forethought were central tenets of the celebrated cluster of shared Victorian values that had helped to make Lancashire and Britain the 'workshop of the world'.[105] Furthermore, terms such as 'improvident', 'careless', and 'short-sighted' were part of a vocabulary more commonly employed by the Victorian middle classes to berate the 'feckless' and 'spendthrift' poor. Put in perspective, it is very likely that the campaigners' stinging rebukes discomfited more than a few of the city's polluting industrialists.

When the *Manchester Guardian* leader comments on the loss that the smoke-cloud represented to the city's businessmen, it is almost certainly referring to a pecuniary loss. From the outset, this thread of the story line placed great emphasis on appeals to economic rationality. It was postulated that factory owners could achieve significant savings on coal bills through installing furnaces constructed to consume their own smoke, or by ensuring that their stokers were well-trained men who took proper care in performing their tasks. The Select Committee on Steam Engines and Furnaces of 1819 was informed that one new smoke-consuming device could cut expenditure on fuel bills by as much as 75 per cent, although later Victorian claims regarding annual savings were generally more plausible, fluctuating between 5 and 25 per cent.[106] However, it must be noted that early smoke abatement technology was by no means reliable, and could actually result in more fuel, rather than less, being burnt.[107] Nonetheless, reformers never tired of reminding manufacturers about the potential savings to be made from the careful husbanding of valuable fuel resources, while at the same time drawing attention to the costly damage that air pollution caused to private property and household goods.

This strategy unquestionably looked to Edwin Chadwick's successful use of fiscal arguments in swinging public opinion behind his campaign for a comprehensive programme of sanitary reform.[108] Chadwick and other public health reformers continually stressed the great financial burden to the British taxpayer of preventable illness caused by polluted water supplies and woefully inadequate urban sewerage and drainage systems. Yet, initially, it appears that the harm that coal smoke caused to the people's health played no more than a supporting part in the narrative of waste.[109] An early petition to the House of Commons against smoke pollution, organised in 1843 by the Manchester Association for the Prevention of Smoke, does not explicitly mention concerns over health. Instead it points to the needless waste of fuel resources by manufacturing industry, and represents the destructive results of 'excessive smoke' on material possessions as nothing less than a burdensome local tax.[110] To begin with, the campaigners were to focus mainly on promoting new technological means of making more economical use of coal and directing public attention to the draining effect that smoke pollution had on 'the weekly expenses of every family residing within its influence'.[111] In 1842 Henry Houldsworth, industrialist and prominent member of the Association, calculated the financial cost to Manchester's inhabitants in 'washing, cleansing, and keeping clean persons, garments, furniture and houses' to be 'not less than £100,000 per annum'.[112] It was one of the first cost-benefit exercises undertaken by Victorian anti-smoke activists in trying to persuade people to reduce air pollution in urban areas – but by no means the last. For example, at mid-century Chadwick himself had estimated that coal smoke was costing the inhabitants of London no less than £3 million every year in wasted fuel and excessively high washing bills.[113] The idea that in one way or another air pollution cost the whole community a considerable amount of money performed the dominant discursive role during this early period of anti-smoke activity.

After 1850 the notion that the prevention of smoke would 'frequently result in ... pecuniary advantage' to both householders and businessmen continued to be widely disseminated.[114] At London the meteorologist Rollo Russell listed some twenty-four different forms of loss or damage caused by smoke in his cost-benefit exercise of 1889. These included: the extra gas used for lighting all year round due to loss of sunlight; damage caused to the 'delicate' stock of drapers, milliners, and other textile traders; reduced capacity for work due to ill health; and the destruction of trees, plants, shrubs, flowers, vegetables, and fruits. Although Russell did not include 'uncertain items', such as the effects of the residence outside the Metropolis by all those wealthy enough to avoid the smoke, the total cost still reached £5,200,500 per annum.[115] In the early years of the twentieth century Manchester's Air Pollution Advisory Board reported that the city's smoke was costing its householders almost £1,000,000 every year, with no less than £242,705 of this figure arising from extra washing alone.[116] The Board's report continued,

> Not only does black smoke mean waste in itself, but it causes further waste. Everybody knows how much it disfigures, but it is not generally known how much it destroys. It levies what may be called the Black Smoke Tax, and everybody living in Manchester pays this tax. ... Black smoke means not only an aesthetic but also an economic loss.[117]

In addition to demonstrating the great pecuniary burdens that the 'Black Smoke Tax' placed upon the public, reformers continuously depicted coal smoke as the visible failure of manufacturers and householders to capitalise economically on the nation's coal stocks by burning their fuel efficiently. London's smoke-filled air was depicted in 1855 as a vast, unused 'aërial coalfield' by the censorious *Times* newspaper.[118] Two years later the members of Manchester's Literary and Philosophical society were urged to consider industrial air pollution from an 'economical view', as

... of all the immense quantity of coal consumed in manufacturing operations in Manchester, which is estimated at 1,000,000 tons, 416,600 are represented by the water evaporated or work done, while 583,400 tons are represented by the waste that is sent in the form of carbon or carbonic oxide through our large chimneys.[119]

Nearly 60 per cent of the coal burnt in Manchester's mills and factories in 1857 was alleged to be contaminating the city's atmosphere having failed to perform any productive work whatsoever. The inefficiency of Britain's steam power was an oft-recurring motif of the sceptical narrative regarding smoke. In 1890 the conservationist Canon Hardwicke Rawnsley spoke indignantly of the 'dense clouds of unburnt and wasted carbon' he had seen issuing from the tall chimneys of Lancashire's textile mills.[120] The Metropolitan-based Coal Smoke Abatement Society, while estimating that no more than 2 per cent of Britain's coal was lost in the form of soot in 1911, (almost 3,700,000 tons), protested that, 'The frequent emission of dense black smoke is always a sign of avoidable waste. It shows that valuable burning gases are being extinguished and wasted instead of fulfilling their function of turning water into steam.' [121] Smoke control was ceaselessly posited as a valuable business proposition for industrialists, with anti-smoke campaigners claiming that the adoption of improved, mechanised fuel technology, or even 'ordinary care' being taken in the operation of hand-fired furnaces, could mean substantial financial savings.

Unlike the industrial chimney, before mid-century the domestic coal fire was not heavily criticised by contemporaries on the grounds of waste or economy. This situation slowly began to change, however, as alternatives to the traditional open fire became more widely available, such as the 'miserly' coke burning Arnott stove and penny-in-the-slot gas stoves and gas fires. By the 1880s the 'cheerful' domestic hearth was also being regularly denounced by reformers as wasteful and uneconomic, with one of the most disparaging condemnations coming from the eminent scientist Oliver Lodge, Professor of Physics at Liverpool University:

Our universal method of obtaining domestic heat in this country is the open coal fire, and what I want to inveigh against and see altered is this burning of crude coal as it comes from the pit. To shovel a heap of coal together is a simple plan of making a fire – a very simple plan; it is what one would expect of a savage – it would be highly creditable to an ape. To us in the present day it is a disgraceful barbarism ... It is a troublesome process, a wasteful process, a dirty process, and is, really and truly, an expensive process.[122]

Similarly, the *Builder* in 1899 presented the 'pall of smoke' from Britain's domestic fires as symbolic of wasted money, arguing that by the adoption of more efficient fire-grates London alone could save some £2 million in 'hard cash' each year.[123] According to the narrative of waste, clouds of black smoke denoted nothing less than 'a visible and glaring proof' of pounds sterling haemorrhaging needlessly from Britain's industrial and domestic chimneys into the skies above.[124] To the Victorian and Edwardian anti-pollution campaigner coal smoke was indubitably valuable matter in the wrong place, and manufacturers in particular were strongly urged to 'consider it in the same practical commercial spirit which animates them in their other transactions'.[125]

Energy and entropy

Economic concerns over wasted fuel were augmented after mid-century by severe criticisms of the irresponsible depletion of Britain's natural energy resources. In 1850, the German physicist Rudolph Clausius had formulated the second law of thermodynamics, developing the concept that all energy becomes disorganised and dissipates over time, eventually being lost forever.[126] As ideas concerning entropy filtered down to the public domain, industrialists, who treated the nation's cheap and abundant coal as if it was an inexhaustible asset, were increasingly urged by reformers to conserve what was now recognised to be a finite resource. In 1854 Dr Neil Arnott, noted sanitary reformer and designer of the aforementioned fuel-saving stove, vehemently denounced the thriftless squandering of Britain's coal reserves:

> ... coal is a part of our national wealth, of which, whatever is once used can never, like corn or any produce of industry, be renewed or replaced. The coal mines of Britain may truly be regarded as among the most precious possessions of the inhabitants, and without which they could never have attained [such] impor-tance in the world ... To consume coal wastefully or unnecessarily, then, is not merely improvidence, but is a serious crime committed against future genera-tions.[127]

Other contemporary voices took up the new and sobering concept that the profligate depletion of finite energy resources was a 'crime committed against future generations'. For example, John Percy, Professor of Metallurgy at the Royal School of Mines, explained the notion thus to the readers of the *Quarterly Review*:

> Nations, like individuals, when overflowing with wealth, are too apt to be reckless as well as lavish, and to go on scattering their resources broadcast, until they suddenly find themselves ruined beyond hope of redemption. Thus have we sinned with regard to our coal – that matchless reservoir of force – and thus shall we fall from our high estate if we proceed in our mad career of waste and

extravagance ... Our successors will have bitter cause to deplore our folly in this respect and to regard us as spendthrifts, who have ignorantly or knowingly destroyed so much of that force by which we have in great measure achieved our prosperity and our position amongst the nations of the earth.[128]

Percy, along with a growing number of Victorians, clearly understood that energy once consumed was irretrievably spent, and he stressed that later generations of Britons could find themselves severely disadvantaged by the unnecessary waste of limited coal reserves.[129]

By the 1860s there were serious concerns as to just how long coal supplies in Britain would last, with Sir William Armstrong, in his presidential address to the British Association, calculating that stocks would last only for another 212 years. In 1865, the economist W. Stanley Jevons heightened these concerns when he estimated that Britain's coal reserves would be exhausted in no more than a century.[130] The question of a threat to Britain's future power and prosperity was frequently highlighted, and featured prominently in Jevons' influential work. 'When our main-spring is here run down, our fires burnt out', he wrote, 'Britain may contract to her former littleness, and her people be again distinguished for homely and hardy virtues ... rather than for brilliant accomplishments and indomitable power.' [131] Fears over the availability of coal supplies, and the lack of economy in its consumption, were in part allayed by the findings of a royal commission into the coal question in 1871. The noted statistician Price Williams countered the 'alarmist computations' of Jevons by fixing the date of exhaustion of Britain's 'coal cellars' at 360 years in the final report.[132] Nonetheless, the seeds of doubt had been sown, and anti-smoke reformers continued to urge manufacturers to conserve fossil fuel as a 'preventive measure', as 'the Report of the Commission is only another way of saying scientifically that we know very little of what is going to happen'.[133] Without the conservation of coal, a major source of industrial and imperial power, it was argued that Britain had nothing to look forward to but a gradual decline into mediocrity.

Ideas that connected smoke with wasted natural resources were commonplace in the popular press of Britain in the last third of the century. In 1889 a *Manchester Guardian* account of a crowded public meeting against smoke pollution, held at Manchester Town Hall, voiced the main themes of this dimension of the narrative. Lord Egerton of Tatton, who chaired the meeting, claimed that the 'consumption of smoke' would result in 'pecuniary gain' to the city's manufacturers, who needed to be 'enlighten[ed] ... as to their own interests, to show them that the emission of smoke into the air was a waste of valuable carbon which ought to be in the furnace'. While Sir William

Houldsworth, MP for Manchester Northwest, told the large and influential gathering, 'He believed himself that the consumption and prevention of smoke was economical ... because the black smoke which they wanted to stop was actual force and energy going into the air.' [134] The *Gentleman's Magazine*, in an article discussing the possible exhaustion of Britain's coal supplies, urged contemporaries to think of smoke as 'heat and power in the wrong place'.[135] Smoke pollution was increasingly vilified as 'barbarous and unscientific', and readers of newspapers and magazines were regularly bombarded by representations of the belching chimney as indicative of wasted money, energy, and natural resources. In the early years of the twentieth century the *Spectator* spoke of the smoke cloud hanging over London, Manchester and their 'sister cities' as 'the flag and standard of the Age of Dirty Waste in which we live'.[136] And while the 'scandalous waste' of coal stocks was always a major facet of this sceptical story line, other fundamental themes, such as the harm smoke caused to health, also challenged the long-standing association of progress with pollution.

Smoke, the destroyer of health

The damage smoke caused to the health of Manchester's inhabitants was an important component of the narrative of waste and one that acquired an increasingly apocalyptic edge as the century progressed. The detrimental effects of coal smoke on the human respiratory system had been pointed out as early as 1659 by John Evelyn in his pamphlet *A character of England*, where he observed that 'pestilent[ial] smoak ... so fatally seiz[es] on the lungs of the inhabitants [of London], that the cough and the consumption spares no man'.[137] However, at Manchester concerns about the smoky atmospheric conditions causing ill health did not surface conspicuously until 1842, when the Manchester Association for the Prevention of Smoke was formed. The Reverend John Molesworth, the Association's chairman, told the Select Committee on Smoke Prevention of 1843 that the smoke-drenched air of the city was 'no doubt unhealthy' and 'must tend to disease'.[138] Soon after, a comprehensive report on the causes of death in Manchester for the three years 1840-2 seemed to confirm this notion. Compiled for the eminent sanitary reformer Lyon Playfair by John Roberton, surgeon to the Manchester Lying-in Hospital and mainstay of the Manchester Statistical Society, the report showed that 'diseases of the windpipe and chest' were far and away the city's major killers, accounting for almost 30 per cent of all fatalities, or 8,352 of 28,324 deaths.[139] Manchester's smoke-filled atmosphere, Roberton declared, was undoubtedly a significant factor in causing respiratory disease, having a marked 'debilitating influence ... on the health of all, but especially of the young'.[140] While not at first so prominent a theme as the waste of money, fuel and energy, the unnecessary destruction of life and health caused by acidic smoke pollution, often couched in statistical terms, gradually became a vital and frequently utilised strand of the story-line.

It is difficult to substantiate direct links between air pollution and ill health as chronic respiratory diseases have a complex aetiology and because exposures to harmful pollutants vary considerably and are not easy to quantify. Nonetheless, by systematically collecting statistical information on death rates from lung diseases in Manchester, and then comparing the results with corresponding mortality data drawn from less contaminated British towns and cities, anti-pollution activists hoped to prove conclusively the connection between coal smoke and a growing incidence of serious respiratory problems in the city.[141] The stark statistical evidence would tell a story that the pro-smokers could not easily gainsay, adding a further powerful dimension to the narrative of waste and inefficiency by speaking incontrovertibly of the needless loss of human life. By conveying their message in the 'objective' language of mathematics the reformers intentionally exploited a well-established technique for signposting social ills. Indeed, statistical inquiries into the condition of the lower classes were often seen as a 'necessary preliminary' to effective government intervention and action.[142] In Victorian Britain 'scientifically' collected numerical data were commonly treated as 'irrefutable facts', providing the basis for important reforms in public health, education, and shortening working hours.[143] Set against this backdrop, it is not surprising that anti-smoke campaigners frequently used statistical information in an attempt to mobilise public opinion and the government against air pollution.

Statistical studies of common respiratory diseases undertaken by pressure groups, Medical Officers of Health and the office of the Registrar-General repeatedly pointed to Manchester and Lancashire's prodigious death rates from bronchitis and pneumonia. But not all explicitly or exclusively refer to smoke pollution as the direct or indirect cause of numerous premature deaths. However, many such surveys did attempt to show the adverse effects that acidic smoke emissions had on the health of the general public. Manchester's Medical Officer of Health, John Leigh, in a report on coal smoke to the Health and Nuisance Committees of the City Council, found that as well as acting directly as an 'irritant to the respiratory organs', air pollution also had a 'depress[ing] and enervat[ing]' effect on the human constitution, making for the 'ready reception of disease'.[144] In the course of his inquiry of 1883, Leigh compiled the following table of deaths from respiratory diseases, based on the Registrar-General's latest figures, which, he pronounced, 'stood out ... in all their black and hideous simplicity'.[145]

Cheshire and Lancashire	4381 per million persons
London	4365 per million persons
Yorkshire	3890 per million persons
Monmouth and Wales	3551 per million persons
Essex, Suffolk, Norfolk	2979 per million persons
Surrey, Kent, Sussex, Hants, Berks	2835 per million persons

Table 6. Deaths from Respiratory Diseases in England and Wales.

Source: John Leigh, 'Coal Smoke: Report to the Health and Nuisance Committees of the Corporation of Manchester', *Health Journal and Record of Sanitary Engineering*, Vol.1, 1883.

According to Leigh, other significant factors that caused chronic lung complaints, such as dilapidated housing and freezing climatic conditions, could not account for the contrasting experience of industrial and rural areas that these statistics so plainly revealed.[146] The 'dirty, clumsy, wasteful unscientific' combustion of coal in Manchester and Lancashire was undeniably the source of many of these unwarranted 'excess' deaths.[147] In 1882 Dr Arthur Ransome, a nationally respected authority on respiratory ailments, calculated that in the preceding decade some 34,000 people had died from 'diseases of the lungs' in Manchester and Salford. Ransome argued that smoke pollution was a significant – and preventable – cause of this mortality and that its abatement 'would save many useful lives in the next generation'.[148] The author of an earlier report published by the Manchester and Salford Sanitary Association had concluded that because of its polluted atmosphere, 'Manchester suffers more from diseases of respiratory organs than any town or city in England'.[149] Similar surveys were conducted by anti-smoke societies elsewhere in Britain. For example, in the early years of the twentieth century the London-based Coal Smoke Abatement Society utilised figures showing that the smoke-filled fogs of Glasgow and the Metropolis caused more destruction of life each year 'than occurred in any battle during the Boer War'.[150] Then, as now, reformers consistently used the burgeoning death rates from respiratory diseases to keep the issue of air pollution in the public eye.

Concerns about the health effects of coal smoke on children living in urban-industrial areas made up a notable part of this component of the narrative. The harm smoke pollution caused to the young was a potent motif, as mortality rates for infants and children aged between one and five remained stubbornly high during the period.[151] And, as previously stated in part one of this study, in England and Wales bronchitis and pneumonia were the main killers of those under five in the decades from 1860 to 1900, while wasting diseases and diarrhoea were the principal causes of death for infants under three months of age. In 1843 John Roberton's report on the causes of death in Manchester emphasised the adverse effects that smoke had upon the health of young children:

> The atmosphere, sluggish and polluted in certain districts where the population is dense, and impregnated with smoke in all, is unquestionably a cause of juvenile death. Its effects are most pernicious. By weakening the vigour of the system the infant is rendered sickly, and strongly predisposed to disease, and surrounded by so many morbific causes, is often speedily cut off. Families who can afford have mostly removed beyond the smoke ... [152]

In his condemnation of the 'constant dark canopy' of smoke that shrouded Manchester's streets, John Leigh particularly stressed the death toll on the city's children, who 'contributed nearly one-half of the mortality of the district before they attained the age of five years'.[153] The vital statistics collected by Victorian social investigators showed clearly that urban children had a much poorer chance of survival than their rural counterparts, even if born to a comfortably-off middle-class family.[154] Not surprisingly, the upper and professional classes of the northern factory towns usually chose to raise their families in the cleaner, healthier air of the surrounding Cheshire countryside.

The 'objective' use of columns of figures and dry statistics was regularly leavened by emotive references to air pollution's cost in human lives, especially the lives of 'helpless and innocent' urban children.[155] After a prolonged smoke-laden fog had engulfed Glasgow in November 1909, the city's death rate jumped alarmingly. In the space of just two weeks mortality increased from 18.0 per thousand to the annual equivalent of 32.7 per thousand. It was an acceleration in fatalities from respiratory diseases that fuelled this surge, with this category of mortality rising apace from 25 per cent of deaths from all causes to a peak of 40 per cent.[156] The figures were believed to demonstrate the 'indisputable fact' that smoke pollution had caused well over 500 'excess' deaths in Glasgow, mainly from bronchitis, pneumonia and pleurisy.[157] Although mortality among the elderly and the sick was by no means played down, anti-smoke campaigners claimed that it was the very young who had suffered most in the dense acidic fog. In a public lecture delivered at Glasgow's Technical College, Peter Fyfe, Chief Sanitary Inspector, bemoaned the fact that in 1909 of the 3,059 children who died in the city before reaching their first birthday, some 667 of these infants had succumbed to smoke-related respiratory diseases.[158] Fyfe unequivocally stated that Glasgow's 'modern Erebus – the vomiting chimney' was the root cause of their untimely deaths:

Let us use no euphemisms, no glossing words, to cover our own misdemeanours in the vain attempt to blame Dame Nature. The citizens themselves, along with some manufacturers, are alone to blame; and the dire effect, death – this excessive death – is due to one thing and one alone, and that is smoke! And the pity of it all is that it is mostly our children that are slain.[159]

William Brend, Lecturer on Forensic Medicine at Charing Cross Hospital, London, was equally emphatic with regard to this aspect of the subject. 'Dirtiness of the air', he wrote, 'appears to be the one constant accompaniment of a high infant mortality ... a smoky and dusty atmosphere as a cause of infant mortality far transcends all other influences.'[160] Cautionary statistics portraying a strong association between high death rates and smoke pollution in Victorian and Edwardian Britain repeatedly entered into the narrative of waste and inefficiency. However, the significance of the numerical data amassed by doctors and scientific experts was not uncontested.

While the majority of physicians appeared to accept that coal smoke was a direct cause of spiralling urban mortality rates from chronic respiratory diseases, and an indirect source of all manner of ill-health besides, even at the end of the nineteenth century the issue was not settled beyond all doubt. One dissenter, Harvey Littlejohn, Medical Officer of Health for Sheffield, encapsulated the views of more than a few contemporaries when in 1897 he observed,

We constantly hear it stated that visible smoke is the cause of high death rates, and has a serious influence on the health of the community. In my opinion this has still to be proved. That a smoke-laden atmosphere very considerably interferes with the comfort and thrift of the inhabitants, that it to a varying extent, depending upon certain climatic conditions, has an influence upon the general vitality of a portion of the population, and even acts detrimentally on a small proportion, affecting their health I must admit, but that black smoke has a directly prejudicial influence upon health has not been conclusively proved.[161]

Escalating death rates from lung diseases were also commonly associated with other environmental sources of illness, such as cold, damp weather and badly ventilated, overcrowded housing. Although the bronchitis group of respiratory diseases had unobtrusively become the commonest cause of death in England's towns and cities, consistently killing between 50,000 and 70,000 persons per annum between 1875–94 (Figure 2), it was not easy to establish just how telling a role air pollution played in this mortality. Therefore, while statistical inquiries had led to considerable progress in improving sanitation systems and lowering death rates from epidemic 'dirt diseases' such as cholera and typhoid during the nineteenth century, numerical evidence could not conclusively prove the relationship between a lingering death from acute respiratory diseases and the deepening smoke-cloud.

The Victorians' 'romantic glorification' of those suffering from serious respiratory diseases was to hamper initiatives to improve health in this respect. The sentimental treatment of the dying victims of chronic lung complaints, who were often represented as the 'hero of the moment', differed notably from horror-struck attitudes towards outbreaks of cholera and fever.[162] If Manchester was the 'shock city' of the Industrial Revolution, cholera was undeniably the 'shock disease'. Cholera first struck England in 1831/2, returning in 1848/9, and, with decreasing virulence, in 1853/4 and 1866/7.[163] Contracted by swallowing infected food or water, the disease could kill within a few hours of the first symptoms appearing. In stark contrast to the dignified and courageous suffering of the consumptive or bronchitic person, the symptoms exhibited by a cholera victim were extremely distressing to modest Victorian sensibility, involving a massive loss of body fluids from violent bouts of vomiting and diarrhoea.[164] In the case of cholera, alarming epidemic death rates were reinforced by the dread-inspiring effect the disease had upon contemporaries. Although a smoky environment raised mortality significantly in urban-industrial districts, it did not have the same dramatic impact on the Victorian psyche.

Frustration at the ineffectiveness of 'reform by numbers' in the case of air pollution saw a growing number of impassioned appeals against environmental injustice, with, for example, the socialist Edward Carpenter writing emotively:

> ... any one who has witnessed ... the smoke resting over such towns as Sheffield or Manchester on a calm fine day – the hideous black impenetrable cloud blotting out the sunlight, in which the very birds cease to sing, – will have wondered how it was possible for human beings to live under such conditions. It is probable, in fact, that they do not live ... The workers, producers of the nation's riches, dying by thousands and thousands, choked in the reek of their own toil ... [165]

By the last quarter of the nineteenth century the notion that smoke was contributing markedly to ill health and mortality in the industrial city had reached beyond medical circles to acquire a wider resonance. The idea had become an important part of public discourse concerning urban air pollution, as this letter to the *Manchester Guardian* illustrates:

> ... the great city of Manchester stands in the unenviable position of being one of the unhealthiest cities in the kingdom, with an appalling death-rate behind it. The evils are not far to seek. They are, in my opinion, mainly due to ... the smoke demon and noxious vapours. So much has been written and said on this nuisance that I can add nothing fresh, only to hope that they who are charged with guarding the public health will be alive to their duties and enforce their power

to abate the evil, which is killing downright our boys and girls – the men and women of the future, – and let them breathe pure air and not poison ... it is high time that something practical were done to make it what it ought to be – the second city in the Empire, healthy and cheerful, instead of what it is insanitary and gloomy.[166]

The smoke pollution that veiled Manchester and other factory towns, rather than being a cause for celebration, was represented to the public as nothing less than a funereal pall. By the latter decades of the century, however, concerns about the putative effects of smoke on health were not limited solely to an increase in death-dealing respiratory diseases.

Illustration 7. The 'Smoke Demon'.

Source: *Idler*, Vol.II, 1893.

Smoke and degeneration

Manchester's smoke-laden atmosphere obscured the sun, and this 'destruction of daylight' led to fears that smoke pollution was contributing to the general physical and moral deterioration of the urban workforce – a much debated concern throughout the nineteenth century. As early as 1876, Dr Thomas Andrews, president of the British Association for the Advancement of Science, stated, 'There can be no doubt that the prevalence of smoke in the atmosphere of our large towns tends to deteriorate the physical condition of our people.' [167] In the 1880s, as Britain's position as the world's leading economic, industrial, and imperial power came under pressure, disquiet about the physical inefficiency of the nation gained ground. Campaigners attracting notice to this component of the narrative of waste regularly used imagery of deteriorating, etiolated vegetation as a commonsense analogy to illustrate what they thought was happening to the urban populace. For example, in an article entitled *Smoke, and its Effects on Health*, the pseudonymous *Lucretia* commented brusquely:

> If we place a plant in a dark gloomy cellar, it is no matter of surprise to us if it seems blanched and sickly. We know we cannot expect anything more from it; we have placed it in artificial conditions, and the results are a moral certainty. Yet such is our inconsistency that we express surprise at the stunted forms and complete physical deterioration of the inhabitants of the smoky, lurid back streets of Manchester.[168]

At the turn of the twentieth century, a *Manchester Guardian* leader opined that the 'future of the race' depended to a large degree on the restoration of natural light to the murky city centres of Britain. The *Guardian*, under the editorship of C.P. Scott, a subscriber to the Smoke Abatement League, reminded its readers: 'Our pride in an Empire on which the sun never sets should be tempered by the reflection that there are courts and slums at home on which the sun never rises ... Not what a nation has makes it great, but what its citizens are.' [169] With around a third of the British population living in big, gloomy cities such as Manchester and London by 1900, the question of physical degeneration began to excite the public's imagination.[170]

Victorian city-dwellers did tend to be small in stature, slightly built, and pale in complexion. To be sure, inadequate working-class diets and unwholesome living and working conditions adversely affected the robust physical development of many underprivileged urbanites. However, the absence of sunlight in urban-industrial areas was widely associated with 'the pallid and unhealthy and stunted appearance of our town populations'.[171] What is more, at Manchester in 1882 it was argued that the declining physical efficiency of its poorer inhabitants, caused directly by the leaden, polluted skies, threatened the very survival of the city itself:

> A stunted, scrofulous, and ricketty working population has been raised up in our midst ... The present existing facts of physical and moral deterioration of the human type which apparently lives, eats, and has its being amongst the lower

classes of this large city, certainly shakes to its foundation Darwin's theory of the 'survival of the fittest' ... the smoky atmosphere of Manchester is not *all* that can be desired ... The clearing of the atmosphere is one of the greatest necessities of the age; and we should consider it so ... the proof of the great dangers existing to the prosperity of Manchester, socially, physically, and commercially, is the apathy of the general mass of its inhabitants with regard to the 'smoke nuisance'.[172]

As the century wore on, many critics of Manchester's smoke came to agree that the degeneration of the urban poor would not only damage the city, but also Britain's standing in the world. This strand of the story line gained currency in 1899, when of eleven thousand men in Manchester who volunteered for military service in the Boer War, eight thousand 'were found to be physically unfit to carry a rifle and stand the fatigues of discipline'. Concern did not end there, however, as of the three thousand who were accepted to serve in the army, only twelve hundred were found to be even 'moderately' fit.[173] As a consequence of the war in South Africa, during which Britain's urban-raised soldiers were thought to have performed badly in comparison with their country-bred opponents, serious doubts were raised over the whole future of empire.[174] In 1902 the Bishop of Manchester, Dr Moorhouse, addressing the Jubilee Conference of the Manchester and Salford Sanitary Association (at which the conservation of sunlight was a prominent theme), commented:

> The Boers, though poor farmers only, few in number, and with little knowledge or culture, were a rich people – 'rich in manly and physical qualities.' No race in the world stood in greater need of these high moral and bodily qualities than our own. Our Empire had undertaken a task of mighty responsibility, and its due discharge would depend on the virility and vitality of our people.[175]

Only a few months later, Fred Scott, the secretary of the Manchester and Salford branch of the Smoke Abatement League, warned that the 'smoke nuisance' was actively undermining the health and vigour of the nation. Polluted city air, Scott contended, was in no small part responsible for 'sapping the virility of our people – that grand heritage which has made the British the greatest colonising and conquering race the world has known'.[176]

The popular newspapers of late-Victorian and early-Edwardian Britain frequently reflected serious reservations about the security of the empire, as its rivals grew ever bolder.[177] Rising great powers such as Germany and the United States, with impressive rates of population growth, and reputedly breeding healthier workers and soldiers, were expanding their interests across the globe. And it was by no means certain to contemporaries that Britain had the requisite human resources to adequately protect its overseas territories from their advances.[178] The ethos of imperialism was based on virility: thus in an increasingly hostile international climate it was argued that Britain 'require[d] as its first condition an Imperial Race ... [as] the survival of the fittest is an absolute truth in the conditions of the modern world'.[179] Anti-pollution activists capitalised on these growing anxieties by claiming that coal smoke, in combination with other factors such as dietary deficiencies, endangered Britain's global primacy by impairing the development of 'a manly, vigorous, enterprising, healthy race which will hold its own against all foreign competition'.[180] The colonies were depicted as healthy limbs on a decaying body, and the wisdom of the state in continuing to allow the air to be polluted by coal smoke was seriously questioned.[181] From the 1880s on, the reformers' narrative repeatedly emphasised the degeneration of the lower classes of the city, aiming to shake the public's confidence in a viable future for Manchester, Britain, and empire.

Late-Victorian worries about the physical degeneration of Britain's city-dwellers were exacerbated by the notion that the gloomy urban environment was also adversely affecting the 'mental and moral health' of the nation. The murky half-light of the factory towns was thought to have a 'most depressing influence on the mind', which could cause physical lassitude and an inability or unwillingness to produce a good day's work.[182] The idea that impaired mental efficiency might endanger the economic prosperity of the nation was made much of by some activists at a time when Britain was seeking to increase the quality and quantity of its manufacturing output to avoid being eclipsed by its rivals.[183] Moreover, Ralph Carr Ellison, in an address to the Sanitary Institute of Great Britain in 1882, highlighted the way in which air pollution damaged the self-esteem of the lower classes:

> One of the worst effects of unregulated excess in the discharge of smoke from a host of tall chimneys is the perpetual state of unsightly dirtiness which it compels the people of the working class to live in. They can scarcely remain personally clean, or enjoy the honest pride of a clean shirt, except for a very few hours. The consequence is an inevitable loss of self-respect, which is much to be deplored.[184]

Charged as it was with 'sulphurous fumes and with bitter black carbonic dirt', Ellison was also convinced that the polluted atmosphere produced a 'raging thirst for beer, ale, and ardent spirits' among the nation's townspeople.[185] A theme taken up by Manchester's Thomas Horsfall, a leading philanthropist and anti-smoke campaigner, who bluntly informed the city's Statistical Society:

> ... we are living in a city in which it is admitted that a considerable proportion of the people are drinking themselves to death, and, what is much worse even than that, are drinking their innocent wives and children into pauperism and wretchedness; when one remembers that a very large proportion of the inhabit-

ants have become the slaves of betting and other forms of gambling; that there are thousands of people in the town whose hearts and brains are atrophied by the lack of wholesome knowledge; and that there are many other evils, partly at least *due to the nature of the environment of the people.*[186]

Horsfall, together with many other Victorians, believed that the two main environmental factors responsible for producing a brutalised, degenerate urban populace were the 'smoke nuisance' and an absence of nature in Britain's factory towns.[187] Reformers claimed that to compensate for the grim local environmental conditions – the lack of daylight, natural colour and vegetation in the drab industrial city – many among the working classes were wasting their scant financial resources on 'vulgar' entertainment, games of chance, and drinking excessively. To get drunk was, in the memorable phrase of Mr Justice Day, 'the shortest way out of Manchester'.[188] In the ultracompetitive age of the new imperialism, Britain's workforce was widely perceived to be deteriorating mentally and morally, as well as physically. The urban poor were accused of lacking reverence for God, high moral character, and a proper sense of duty.[189] Due to the malign influence of the 'smoke demon', countless city-dwellers were assumed to be straying from the straight and narrow path. What is more, high levels of air pollution ensured that there were few representatives of the middle and upper classes resident in Britain's factory towns who could set a 'proper' example for the errant working classes to follow.

From the early Victorian period on, air pollution contributed in no small part to a worrying 'separation of the classes'. To escape the smoke and stinks of the city Manchester's moneyed classes had fled to the cleaner and greener climes of places like Alderley Edge, Altrincham, and Wilmslow. A lack of social contact with their 'betters' was often cited as an explanation for the 'failings' of the masses.[190] Writing in 1843, John Roberton emphasised that the unravelling of traditional ties had deprived the labouring masses congregated in Manchester and other industrial towns of much needed moral and spiritual guidance:

> It is a misfortune almost peculiar to their lot that they live, in a measure, by themselves, few mingling with them as residents in their particular localities, to whom they can look up as their superiors in station and intelligence. In the township of Manchester so very large a proportion of the families of the better classes, including in general even those of the clergy and other ministers of religion, have withdrawn from it to a purer air, that the inhabitants are almost exclusively labourers. The consequence is that there remain only very few, in some districts possibly not one, who by education and position in life, are fitted to exhibit an improving example or to perform those numberless acts of kindness and benevolence, so often and so greatly needed in such a population.[191]

By the 1880s, the middle-class exodus from the smoky environs of Manchester had been almost completely effected: so much so that John Leigh described Manchester as being 'from a residential point of view, a city of cottages'. Moreover, this 'estrangement' between the classes was commonly believed to be 'productive of many evils'.[192] Potential civil discord, drinking, gambling, crime, and prostitution were just some of the various 'evils' that social and environmental reformers warned against.

Following the Boer War, anxiety over declining national efficiency saw the government set up an Inter-Departmental Committee on Physical Deterioration to examine the effects of factors such as alcoholism, insufficient food, overcrowding, and environmental pollution on Britain's city dwellers. Coal smoke featured prominently in its report of 1904, with the Committee airing many of the key issues relating to this dimension of the narrative of waste. Thomas Horsfall called the Committee's attention to the 'careless' pollution of Manchester's atmosphere by coal smoke, arguing that it was 'the cause of much disease and physical deterioration' in the city.[193] Furthermore, he informed the inquiry that Manchester's billowing chimneys had led to 'the removal of all well-to-do persons from the town', which in turn was 'a most fruitful cause of the ignorance and bad habits of the poor'.[194] The Committee strongly recommended that tougher legislative action should be taken against industrial coal smoke, while 'ordinary householders' were to be made more aware of the role that they played in blackening the skies.[195] Smoke pollution was a serious cause for concern in a world of growing imperialistic rivalries. By the turn of the twentieth century there was a widespread belief that only the strongest and most intelligent nation would come out on top in the global 'struggle for existence'. Manchester's anti-pollution campaigners argued that until smoke was abated, and traditional social bonds were re-established in the city, numerous charitable initiatives aimed at restoring health and 'raising the level of civilisation' among the 'degenerate' poor were doomed to failure.[196] In short, much needed moral, intellectual, and physical improvement of the urban workforce would not be forthcoming without there first being a radical transformation of the unwholesome atmospheric conditions.

The 'unloveliness' of Manchester

The cityscape of early Victorian Manchester, dominated by its factories and tall, smoking chimneys, had captured the world's imagination. The dense cloud of air pollution that shrouded the city was instrumental in creating its enduring identity as a 'mighty region of smoke and toil'.[197] However, while the smoke-blackened fabric of the Manchester skyline may have stood for enterprise, wealth and well-being in the eyes of many observers, it symbolised the very antithesis of 'progress' for others. Trapped in barren, monochrome streets, the northern city dweller was often portrayed as living in a world bereft of the pleasing surroundings, 'high' culture and urbane society generally associated with older world cities. In 1844 Leon Faucher's first impressions of industrial Manchester were 'far from favourable':

> Its position is devoid of picturesque relief, and the horizon of clearness. Amid the fogs which exhale from this marshy district, and the clouds of smoke vomited forth from the numberless chimneys, Labour presents a mysterious activity, somewhat akin to the subterraneous action of a volcano. There are no great boulevards or heights to aid the eye in measuring the vast extent of surface which it occupies. It is distinguished neither by those contrasting features which mark out the cities of the middle ages, nor by that regularity which characterises the capitals of recent formation. All the houses, all the streets, resemble each other; and yet this uniformity is in the midst of confusion ... The waters of the Irk, [are] black and fetid ... [there is a] want of public squares, fountains, trees, prom-enades, and well-ventilated buildings; but it is certain that it would be a difficult task to devise a plan by which the various products of Industry could be more concentrated ... You hear nothing but the breathing of the vast machines, sending forth fire and smoke through their tall chimneys, and offering up to the heavens, as it were in token of homage, the sighs of that Labour which God has imposed upon man ... at Manchester, industry has found no previous occupant, and knows nothing but itself. Every thing is alike, and every thing is new; there is nothing but masters and operatives. Science, which is so often developed by the progress of industry has fixed itself in Lancashire. Manchester has a Statistical Society; and chemistry is held in honour; but literature and the arts are a dead letter. The theatre does nothing to purify and elevate the taste, and furnishes little but what is necessary to attract the crowd habituated to gross pursuits ... Everything is measured in its results by the standard of utility; and if the BEAUTIFUL, the GREAT, and the NOBLE, ever take root in Manchester, they will be developed in accordance with this standard.[198]

Anxious to transform its boorish, unflattering image, after mid-century Man-chester attempted to 'reinvent itself' in architectural terms as a fair 'medieval city'.[199] The city's 'commercial palaces' were designed in the elegant *palazzo* style of Renaissance Italy, while the naturalistic neo-Gothic style predominated in monumental public buildings such as Alfred Waterhouse's Town Hall, constructed over nine years at a cost of £1 million. The ongoing 'beautification' of the city also embraced the laying out of new public parks that, like its neo-Gothic architecture, were designed in part to bring a semblance of nature back into the polluted city to 'uplift and civilise' its working masses. Moreover, the city's built environment also conveyed another important message. Manchester's innovative architecture was designed to reflect the sophistication, power and status of its well-to-do businessmen and professionals to the wider world. To paraphrase John Seed, if parvenu Manchester was to be respected as a major international city it needed to proclaim its status not only as a centre of industry and commerce, but also as a place of culture and refinement worthy of being spoken of in the same breath as Athens, Florence, London, Paris or Rome.[200]

The architecture of Manchester, however, rapidly began to blacken, and its elaborately carved stonework to crumble, due to the corrosive action of the viscid, sulphurous smoke-cloud. Thomas Horsfall questioned the wisdom of the city council in spending a million pounds on a town hall, when those who commissioned it must have known that it would be 'ruined by soot and made nearly invisible by smoke'.[201] Furthermore, as part one of this study established, trees, shrubs, and flowers all struggled for survival in Manchester's newly opened parks due to the problem of acid rain. By the latter decades of the nineteenth century the 'unloveliness' of Manchester was a sensitive issue for many contemporaries, since from an aesthetic perspective the 'great city of smoke and toil' was thought to compare woefully with the older world cities it had wished to rival. What is more, even after the construction of the city's impressive public libraries, museums and art galleries, the myth that the whey-faced 'Manchester Man' was something of a cultural lowbrow, interested only in business, balance sheets and matters of fact persisted.[202] In 1877, for example, John Ruskin famously pronounced that Manchester could produce 'no good art, and no good literature'; a stinging rebuke for a city that had invested so heavily in the 'finer arts'.[203] Despite undertaking several decades of expensive 'self-glorification and self-improvement', high levels of smoke pollution meant 'Cottonopolis' had largely failed to re-position itself as an eye-catching, cosmopolitan city of note.[204] Anti-smoke activists complained that rather than being thought of as 'the second great city in the Empire' after London, Victorian Manchester's name had instead become 'synonymous with everything that is depressing, dingy-looking, and smoke begrimed'.[205] A *Punch* cartoon of 1882, that poked fun at the notion of the planned Ship Canal bringing the 'seaside' to Manchester, lampoons the city's grimy image (Illustration 8), while the socialist painter and illustrator Walter Crane similarly portrayed Manchester as 'The Cinder Heap' (Illustration 9). Notwithstanding the erection of numerous architectural works of art, Manchester was still widely considered to be 'uncultured' and 'uglier than nearly all other towns'.[206]

Illustration 8. Manchester-sur-Mer.

Source: *Punch,* 1882.

In the north of England during the 1890s the emerging socialist movement also emphasised the ugliness and wastefulness associated with Lancashire's smoke-capped chimneys. Contributors to Robert Blatchford's *Clarion* newspaper regularly denounced the 'evils' of smoke pollution, and the way in which industrialisation and urbanisation was turning Britain into 'a dull, sunless cinder-heap'.[207] Founded in 1891, the *Clarion* achieved the widest circulation of any socialist newspaper of the period by selling some 80,000 copies per week to the workers of the northern factory towns.[208] Blatchford's revulsion at the environmental consequences of industrial pollution and the misery of urban life was voiced particularly strongly in his book *Merrie England,* which sold two million copies in the fifteen years after its publication in 1894.[209] The work first appeared as a series of articles in the *Clarion* and repays quotation at some length:

Illustration 9. The Cinder Heap.

Source: C.Rowley, *Fifty Years of Work Without Wages*, 1912.

The Manchester School will tell you that the destiny of this country is to become 'The Workshop of the World.' ... But if this country did become the 'Workshop of the World' it would at the same time become the most horrible and the most miserable country the world has ever known. Let us be practical, and look at the facts. First, as to the question of beauty and pleasantness. You know the factory districts of Lancashire. I ask you is it not true that they are ugly, and dirty, and smoky, and disagreeable? Compare the busy towns of Lancashire, of Stafford-shire, of Durham, and of South Wales, with the country towns of Surrey, Suffolk, and Hants. In the latter counties you will get pure air, bright skies, clear rivers, clean streets, and beautiful fields, woods, and gardens; you will get cattle and streams, and birds and flowers, and you know that all these things are well worth having, and that none of them can exist side by side with the factory system. I know that the Manchester School will tell you that this is mere 'sentiment.' But compare their actions with their words. Do you find the champions of the factory system despising nature, and beauty, and art, and health – except in their speeches and lectures to you? No. You will find these people living as far from the factories as they can get; you will find them spending their long holidays in the most beautiful parts of England, Scotland, Ireland, or the Continent. The pleasures they enjoy are denied to you ... they point out to you the value of the 'wages' which the factory system brings you, reminding you that you have carpets on your floors, and pianos in your parlours, and a week's holiday at Blackpool once a year ... And let me ask you is any carpet so beautiful or so pleasant as a carpet of grass and daisies? Is the fifth-rate music you play upon your cheap piano as sweet as the songs of the gushing streams and joyous birds? And does one week at a spoiled and vulgar watering-place repay you for fifty-one weeks' toil and smother in a hideous and stinking town? ... Look through any great industrial town in the colliery, the iron, the silk, the cotton, or the woollen industries, and you will find hard work, unhealthy work, vile air, over-crowding, disease, ugliness, drunkenness, and a high death-rate. These are *facts*.[210]

On the cover of the one shilling edition of Blatchford's *Merrie England* trees in leaf and swallows flying in a clear sky represent the beauty of the countryside, while an inferno of darkly smoking chimneys symbolise the degradation of the modern factory town.[211]

Socialists such as Blatchford and Allen Clarke drew heavily upon the romanticised vision of John Ruskin, William Morris, Henry David Thoreau, and other steadfast opponents of urban-industrial society.[212] In order to change for the better the 'smoky lives' of the working classes, the socialists argued that polluting industry should be scaled down to produce only for the needs of small, self-sufficient communities rooted in the earth.[213] In *The Effects of the Factory System*, again originally serialised in the *Clarion*, Clarke wrote:

> I would like to see Lancashire a cluster of small villages and towns, each fixed solid on its own agricultural base, doing its own spinning and weaving; with its theatre, gymnasium, schools, libraries, baths and all things necessary for body and soul.[214]

Utopian socialist notions of a return to nature, of a de-industrialised, smokeless Lancashire where the pursuit of material betterment would be subordinate to an improved quality of life, did, for a short time before the Great War, enjoy some popular support.[215] Moreover, during the last years of the nineteenth century and the early years of the twentieth, there was a considerable degree of overlap between the outlook of the northern socialists and that of the liberal anti-smoke campaigners with regard to the harmful effects of environmental pollution. Both movements utilised and accentuated very similar 'facts' about coal smoke, but for very different ends. Most of Manchester's anti-smoke campaigners simply wanted to clean up the 'workshop of the world' and make it operate more efficiently, they did not wish to see it close down. The reformers asserted that if smoke was abated and something of nature restored to the city, the results would include a happier, healthier, and more civilised urban workforce; an increase in artistic creativity and industrial output; higher standards of production; and, ultimately, the maintenance of Manchester's and Britain's standing in the world. They continually stressed that an investment in the regeneration of the urban environment – in clean air, dirt-free architecture, and flourishing green spaces – was also a way to increase the public wealth of a city. However, despite widespread and favourable press coverage, the narrative of 'waste and inefficiency' failed to overturn the dominant cultural myth that smoke equalled prosperity.

Choosing to live with smoke

By the 1880s, a century's experience of living with smoking chimneys had given a cultural permanence to the notion that coal smoke denoted wealth and well being. The correlation between pollution and prosperity had become so deeply embedded in the stories people told about smoke that many members of northern industrial communities did not often think to complain about the degraded environmental conditions. For example, Allen Clarke wrote of his working-class upbringing in Bolton:

> Living there, I had grown familiar with its ugliness, and familiarity oftener breeds toleration than contempt; I had accepted the drab streets, the smoky skies, the foul river, the mass of mills, the sickly workers, as inevitable and usual – nay, natural, and did not notice them in any probing, critical way.[216]

Indeed, the gloom overhead helped to form, and strengthen, the urban dweller's uncompromising attachment to the 'cheerful' open coal fire. As Professor William Bone observed, 'An Englishman, oppressed as he is [by] ... dreary sunless skies ... seeks relief in his home at nights by his radiant fireside, and disregards with characteristic disrespect the vapourings of scientific cranks who condemn it as wasteful.' [217]

Coal was one of life's necessities as far as the working classes were concerned, and finding a lump of coal in the street was looked upon as a sure sign of good luck. The investigative journalist Henry Mayhew maintained that to the poor of London, 'the importance of coal can be scarcely estimated ... I have found the ill-paid and ill-fed work-people prize warmth almost more than food'.[218] Robert Roberts also considered that obtaining food and warmth were the greatest worries of the slum dwellers of Salford and he noted down the views of a regular customer to his father's shop: "A full belly and a warm backside',

Mrs Carey would announce, 'that's all our lot want! I got a sheep's head boiling on the hob and a hundredweight o' nuts [coal] in the backyard. What more could folks wish for in winter?"[219] The answer to Mrs Carey's question is regular employment, and this is where the potency of the 'wealth' story line really comes into its own. We must not forget that the only occasions on which the labouring poor had experienced clean city air were in times of hardship, such as a trade depression or a strike. Their encounters with a smoke-free urban environment had been uniformly wretched, and under these circumstances it is not difficult to understand why most of Manchester's citizens accepted the polluters' customary story line; they preferred to cling to what they knew to be true. When the city's workers viewed the imposing ranks of smoking industrial chimneys they did see wealth being created, and smoke issuing freely from numerous domestic chimneys did signify their prosperity. Although both main story lines were at times grossly exaggerated, they did carry many different kinds of truth concerning smoke pollution that the city's inhabitants could readily identify with. However, the narrative of 'wealth' was by far the more credible in the eyes of the urban workforce, who knew from bitter, lived experience that a smokeless chimney signified enforced idleness, hunger, and poverty.

The views of the mass of workers have been largely absent from discussions about urban environmental conditions. Working people, however, did hold strong opinions about air pollution, and glimpses of their perceptions of the 'smoke nuisance' in the industrial towns of northern Britain do occasionally come to light – and not only in popular poems and songs. To conclude this part of the book, I would like to suggest that the workers of Manchester employed an evaluative hierarchy where these story lines were concerned.[220] There is ample evidence to indicate that substantial numbers of working class people were worried about the harmful effects of smoke pollution, with, for example, Captain A.W. Sleigh, former Assistant Commissioner of Police at Manchester, telling the Select Committee on Smoke Prevention of 1843:

> The poor people themselves consider it a very great nuisance ... The whole population do. Lord Ashley was at Manchester some time before I left there, and he did me the honour of asking me to go round with him, to look at the condition of the poor people, together with other matters. We visited all the localities minutely, and they all complained of the smoke ...[221]

Ill health, the arduous, Sisyphean task of attempting to keep homes, furnishings and clothes free from soot, and the destruction of vegetation in the barren, smoky city were all issues that attracted their criticism.[222] But the working classes unquestionably afforded a higher priority to the manufacturers' claims that they would be worse off without the industrial smoke with which they had come to associate jobs and economic well-being.

That there was little active support for the smoke abatement movement among the workforce can be exemplified by an account in the *Manchester Guardian* of a public meeting against smoke, convened by the Manchester and Salford Noxious Vapours Abatement Association, at Broughton Town Hall on 11 December 1882. The meeting was planned with a view to putting pressure on Salford Corporation to prevent 'the excessive and needless emission of black smoke ... in the neighbourhood'. The *Manchester Guardian* reported that there 'was a large attendance of working men' at the meeting, and their hostility to the aims of the reformers soon became apparent. Despite an earnest appeal as to the 'vital importance to themselves and their families that their health should be sustained in order that they might continue to earn that income upon which both they and their family depended', the Association's various arguments concerning the deleterious effects of smoke made little impression on the assembled throng. A resolution rebuking the Salford Corporation, on being put to the meeting, 'was lost by a very large majority, only four people in the body of the room voting for it'. Directly after the vote had been taken, Thomas Horsfall, on behalf of the Association, asked, 'Whose voice it was that ordered that no hand be held up in favour of the resolution. Was it that of a manager or overlooker?' The angry crowd immediately shouted Horsfall down. An opposing speaker, George Jones, a member of the Health Committee of the Salford Corporation, found great favour with the audience when he protested that,

> He did not think there was much to complain of in Broughton, and he would be sorry to see a persecution commenced against the manufacturers, for if they were driven from the borough, where would the bread of the working man come from?[223]

The reformers claimed later that, 'It was evident from the opening of the meeting that a large number of workmen who entered the hall ... had been sent to defeat the objects of the Association.' [224] It is impossible to know whether or not manufacturers had coerced their employees into attending the meeting at Broughton. But it is likely that any use of coercion was minimal, as many working men undoubtedly saw their own interests as being intimately linked to those of their employers in this respect.

No single view of air pollution dominated to the exclusion of all others. The story lines of 'wealth and well-being' and 'waste and inefficiency' are two sides of the same coin, with smoke simultaneously existing as both good and evil in the minds of contemporaries in urban-industrial areas. But the strong foothold that the straightforward and cohesive narrative of 'wealth' had obtained in popular culture gave it great influence and staying power. The more fragmentary and scientific story line of 'waste' was not as readily intelligible to the working classes and failed to duplicate or seriously undermine the former's authority. Trust in the purveyors of these differing knowledge claims also influenced the ways that people made decisions about smoke in their uncertain day-to-day lives. Against the backdrop of cyclical economic depressions, especially during the Great Depression years, it is likely that both employers and employees came to share a sense of increased vulnerability and feared change and the unknown. The growing foreign challenge to Britain's commercial supremacy also eroded confidence in Manchester's future as cotton prices fell and production expanded less rapidly.[225] 'Sun doesn't pay hereabout', one worker stated flatly to a reformer in 1890, 'more smoke more work ... at least, that's wot my master says.' [226] The working classes were suspicious of and did not support the reformers' questionable initiatives that might limit industrial growth and endanger their often precarious livelihoods.

Most of Manchester's citizens were not indifferent to smoke, which they had come to view primarily as the 'incense of industry' and only to a subordinate degree as a symbol of waste. Despite the palpability of the coal smoke, this nineteenth-century environmental issue was no less socially constructed or complex than current concerns over intangible air pollutants. The 'waste' narrative attracted widespread public attention but failed to break the hold of the enduring cultural myth that 'wheer there's smook there's brass'.[227] Had they been asked about their goals in life, most working people in Victorian Manchester would have had clear and definite answers: a blazing coal fire in the hearth, a good meal on the table, and the aim of raising – or at least not worsening – their material standard of living. At the end of the nineteenth century, as at the beginning, people still looked to the city's thousands of chimneys to gauge the condition of their world.

Part Three

The Search for Solutions

During the nineteenth century a positive, utilitarian understanding of Manchester's smoke-blackened environment had developed, reflecting the cultural values and beliefs of its fast-growing workforce. In part two of this study I mapped out how contemporaries utilised symbols and stories, which remained compelling over time and predominant in the face of changing circumstances, to shape an image of coal smoke as denoting economic success and social progress. According to the narrative of 'wealth and well-being', Manchester was the 'engine-house' of a thriving and enterprising region celebrated as the 'workshop of the world'. And while the spoils of industry were unequally divided, south-east Lancashire's factory owners and operatives commonly subscribed to the notion that smoke was synonymous with flourishing trade and 'good times'. This interpretation of smoke constituted one of the principal ways in which the industrial communities of cotton Lancashire made sense of their heavily polluted surroundings.

However, by imbuing smoke with a wide range of positive associations Mancunians did not make the everyday realities and degraded conditions of urban life any the less harsh. As the working-class author Allen Clarke pointed out, 'Lancashire is "the workshop of the world" ... but a workshop is generally not a nice place to live in.' [1] Although coal smoke rated a low ranking in a contemporary hierarchy of dangerous atmospheric pollutants, its effects were highly destructive environmentally. And in Victorian Manchester a growing number of contributions to a sceptical alternative narrative began to reflect this fact. Rather than viewing smoke as signifying prosperity and progress, an influential minority of anti-pollution activists represented emissions of sulphurous black smoke as 'barbarous' signs of waste and inefficiency. Even so, few such reformers wished to threaten Manchester's (or the nation's) economic success or future industrial growth in any way. On the contrary, they argued that certain technological fixes, in harness with tighter legislative controls, could greatly reduce the thickening smoke cloud, while saving manufacturers money on their fuel bills and improving their employees' health into the bargain. Many of the reformers' narratives about air pollution, as well as seeking to challenge the beneficial meanings of coal smoke, were at the same time 'problem setting stories'. That is, they were stories that usually contained the blueprint for a proposed programme of remedial action.[2] Yet their pragmatic suggestions to solve the smoke problem met with a lukewarm response from both industrialists and workpeople alike. To better understand why the anti-smoke campaigners' initiatives failed to win enthusiastic public support, part three of this book explores in some depth their search for solutions.

Setting the agenda for change

The 'waste' story line not only helped to frame the problem of the 'smoke nuisance' by giving a coherent (albeit complex) meaning to 'singular and unrelated events' such as dead trees, bronchitic children, and smoking chimneys.[3] It was also used to present new ideas and concepts that set the agenda for change. In 1854, for example, in his lesser known role as journalist, Charles Dickens brought the 'smoke nuisance' to the attention of the many thousands of readers of his successful weekly journal *Household Words*. In his article *Smoke or No Smoke*, Dickens clustered together many of the disparate elements of the narrative of waste, arguing that,

> ... the great destruction of life from pulmonary disease is due to the fact that the soot which smudges the collars and chitterlings of our citizens, that ruins our finest paintings, that blackens our public buildings, that suffocates our country-born babies, that kills our plants, that fleeces our sheep of their whiteness, that blackens our faces, and buries our whole bodies in palls of fog, is also constantly passing into our lungs; and, as the cells of that organ were not intended to act as soot-sifters, any more than Sam Slick's watches were made to be bruised under sledge-hammers, they soon become the 'vile prisons of afflicted breath;' and, stopping it altogether, add mournful entries to the books of the Registrar General of Deaths.[4]

Here Dickens invokes and intertwines several of the narrative threads that by slow degrees came to give an air of permanence to the notion that smoke was indeed 'matter out of place'. But the main thrust of Dickens' article, in common with other 'problem setting stories', is devoted to laying the foundations for a particular course of preventative action.[5] Dickens, concerned that 'Every wreath of smoke that curls up a chimney is so much wasted fuel', for the most part concentrated on promoting the greater use of the latest industrial and domestic smoke abatement technologies.[6] However, in order to elicit the broad sweep of nineteenth-century initiatives to reduce air pollution, rather than focus upon the actions of concerned individuals like Dickens, I now propose to examine more closely the aims and activities of three smoke abatement organisations campaigning in Victorian Manchester. The most significant actors in keeping air pollution in the public eye at this time were Britain's anti-smoke societies, who exerted pressure not only to transform public perceptions regarding the 'smoke nuisance', but also to bring about technological, legislative and government policy changes.

The Manchester Association for the Prevention of Smoke

At Manchester on 26 May 1842 one of Britain's first smoke abatement societies, the Manchester Association for the Prevention of Smoke (hereafter MAPS), was founded. There were several factors that combined to provide the impetus for the formation of this pressure group at this particular time. As the 'Condition of England' question attracted the nation's attention, the 'smoke nuisance' was increasingly targeted by critical sanitary reformers as detrimental to the health and comfort of urban dwellers. In addition, earlier that year a militant anti-smoke meeting held at Leeds, attended by future MAPS chairman the Reverend J.E.N. Molesworth and several other prominent 'Manchester gentlemen', had stimulated interest in smoke abatement not only in 'Cottonopolis' but also in 'the kingdom at large'. Last but not least, the British Association for the Advancement of Science was to meet in smoky Manchester during June 1842 and the city's scientific community wished to bring the subject before them.[7] From the outset, the society focused most of its energy on the search for reliable 'smoke consuming' devices and on investigating techniques to ensure more complete combustion during the 'wasteful' process of raising steam. The presence of Molesworth as chairman of MAPS, and the involvement of other pillars of the church such as the Reverend Richard Parkinson, canon of Manchester, indicates the strength of concern regarding the moral dimensions of the smoke question. However, it was believed that science would provide the solution to the problem, as Molesworth explained to the inaugural meeting of MAPS:

> The difficulty of his position [as chairman] was increased by the consideration that a number of practical men, men of science, and men of business were present, in all of which respects he must acknowledge himself lamentably deficient. He did not make the slightest pretension to either chemical or mechanical science, or to being a man of business; and he felt, therefore, like a dwarf amongst giants, or a child amongst sages. He would endeavour to be the index, the hand of the clock, merely to point to those movements by which the machinery should be actuated: it was by themselves, by men of science and business here, that the subject must be carried out in its details, or it would fall to the ground.[8]

The society's meetings, held in the lecture room of the Royal Victoria Gallery for the Encouragement of Practical Science, were attended by some of Manchester's foremost scientists, technologists, and industrialists, including William Sturgeon, Peter Clare, William Fairbairn, and Henry Houldsworth. The main aims of the new society were set out by Molesworth before the Select Committee on Smoke Prevention of 1843, and despite emerging health concerns not warranting a mention they were fairly comprehensive in their scope:

(1) To establish the fact, by practical evidence, that the excessive smoke in this and other manufacturing districts may in a great measure be prevented by judicious modifications in the construction of furnaces.

(2) To disseminate the knowledge of the principles essential to perfect combustion.

(3) To establish an exhibition of models and drawings of the various improvements by which the production of smoke has been lessened or obviated (with references to parties using the same).

(4) To determine, by experiment, the extent of the saving in fuel by the various plans suggested for preventing smoke.

(5) To induce manufacturers who are favourably disposed to its objects to adopt immediately one or other of the successful plans for preventing smoke; and to provide competent persons to give information to and advise with such parties.

(6) To excite the influence of public opinion against the nuisance by collecting and making known facts illustrative of its evil effects upon the comfort, and its bearing upon the weekly expenses of every family residing within its influence, and ultimately to apply legal means if necessary for abating the nuisance.

(7) To extend the sphere of the society's operations to such towns as contribute adequately to its funds.[9]

Another important object of the society was to maintain links with similar groups formed at Leeds, Huddersfield, London 'and other places – to collect facts ... and form a comparative record as to the extent of the evil'.[10] So how did it happen then, given that the aims of cutting down waste and improving fuel efficiency hardly seem incompatible with the object of money making, that MAPS proved to be an ephemeral organisation which on the whole failed to achieve its practical goals? In order to answer this question, we must first take a closer look at its efforts to improve early smoke abatement technology.

The technological response to smoke

In a nutshell, the reformers believed that the science of efficient combustion and changes in steam technology, which would prove to be of benefit to manufacturers economically, afforded a sure solution to the problem set by the 'smoke nuisance'. However, the anti-smoke activists overlooked or underestimated a whole range of difficulties that meant the desired outcome of technological change – clean air – failed to materialise. Furthermore, the design and implementation of technologies are not simply patterned by an inner 'technical logic', or even by 'economic imperatives'.[11] While conventional, linear models of technological innovation tend to suggest that 'particular paths' of change are in some way inevitable or pre-determined, recent research into the social shaping of technology has demonstrated that many different routes are available. Innovation is a 'garden of forking paths' leading to alternative technological outcomes.[12] The basic message is that artefacts might have been otherwise. When technologies change they are pressed into shape by the interaction of a wide range of technical and social elements:

> The idea of a 'pure' technology is nonsense. Technologies always embody compromise. Politics, economics, theories of the strength of materials, notions about what is beautiful or worthwhile, professional preferences, prejudices and skills, design tools, available raw materials, theories about the behaviour of the natural environment – all of these are thrown into the melting pot whenever an artefact is designed or built ... *all* technologies are shaped by and mirror the complex trade-offs that make up our societies.[13]

It follows, therefore, that in nineteenth-century Britain social, cultural, economic, and environmental factors played as important a role in shaping the development and form of coal-fuelled technologies as did narrowly technical considerations. So the question now becomes, 'What were the specific local concerns and priorities that influenced the development and implementation of smoke-consuming devices in Victorian Manchester?'

The leitmotif of innovation

In the 1840s MAPS endeavoured to induce manufacturers to adopt the latest smoke prevention technology in a city where the predominant world view held that smoke was 'honest dirt', closely associated with both prosperity and jobs. But this was not the only obstacle that the reformers had to face. In terms of the evolution of steam technology, the route that eighteenth and early nineteenth century innovation had taken was to increase gradually the amount of work performed by the steam engine per input of energy, rather than to explore the byway of improved fuel efficiency through pollution control.[14] As the pace of economic growth in Lancashire 'altered from a brisk walk to a businesslike trot', the *leitmotif* of innovation in steam technology was the development of compact, dependable, high-performance engines that yielded more work per pound of coal consumed.[15] By the early nineteenth-century, as economic growth accelerated to a 'full gallop' and adequate access to water power declined, the pressing need to produce an efficient and economical engine that could be widely applied commercially had been met. It was in Manchester that factory owners first went 'Steam Mill Mad', with Lancashire becoming the leading market for the new, improved steam engines produced by Boulton and Watt. However, while the standard of steam-engine technology greatly improved during the eighteenth and early nineteenth centuries, the design and construction of boilers and their furnaces did not keep pace. In 1854 the engineer Charles Wye Williams observed that with regard to innovation in steam power:

> Years were still passing away, and while every other department was fast approaching to perfection, all that belonged to the combustion of fuel – the production of smoke – and the wear and tear of the furnace part of the boiler, remained in the same *status quo* of uncertainty and insufficiency; ... boilers and their furnaces, constructed within the last few years, exhibit greater violations of chemical truths, and a greater departure from the principles on which nature proceeds, than any preceding ones which have come under my observation.[16]

The dominant cultural myth that coal smoke equalled prosperity, and constituted no great threat to life and health, undoubtedly contributed to slowing the rate at which technological change in this second-ranking field of innovation proceeded. But was the question of smoke prevention one so fraught with technical difficulty that Manchester's industrialists were well advised to be reluctant to take the first steps on the path to a smokeless city?

Which path to take?[17]
Although smoke control was very much a secondary goal to increased productivity in the fiercely competitive climate of industrialising Britain, the issue had not been wholly neglected. Indeed, the Select Committee on Smoke Prevention of 1843 heard that there were over sixty smoke consuming devices on the market, which included the addition of some thirty new ones in the previous decade.[18] This labyrinthine situation becomes slightly less confusing when one considers that there were just three main categories of nineteenth-century inventions designed to curb industrial smoke.[19] The first and most important aimed to introduce and mix air into the furnace in order to produce more heat that would, in theory, burn the polluting carbon particles and reduce smoke emissions. This type of device, which aimed to achieve 'perfect combustion', relied heavily on the stoker opening and closing valves or vents to admit jets of air into the furnace during the refuelling process. The most successful example of this class of invention was the 'argand furnace' developed and patented by the aforementioned Charles Wye Williams in 1839. The second category was the more complex 'mechanical stoker', of which the best known were the longitu-

dinal and circular moving grates of John Juckes, patented in 1841 and 1842. This technique also applied the principles of 'perfect combustion', but here the object was to add small quantities of fresh fuel to the fire continuously, so that the ordinary volume of air rising through the grate would be equal to the task of burning the 'unwanted' smoke. The third main category was that of the 'smoke washing' device, a technology that paid little attention to the fire itself. Most inventions of this type utilised flues that were constructed so as to pass coal smoke through sprays of water in order to 'wash out' the tarry soot. Thomas Hedley, a Newcastle ironmaster, patented the most elaborate version of these smoke-washing apparatus in 1842, which, though the least popular method of the day, were the forerunners of the scrubbers and separators used extensively in twentieth-century smoke control.[20] The sheer number and variety of British inventions in this field makes the educational component of MAPS' pragmatic agenda easy to understand.

Despite the best efforts of numerous inventors and Manchester's anti-smoke society, most industrialists insisted that wealth creation and smoke prevention were unrelated activities.[21] And it is certainly fair to say that early smoke abatement technology was in many cases unreliable, as well as being expensive to install and operate. In Manchester, smoke control got off to an inauspicious start at Peter Drinkwater's cotton mill at Piccadilly. During the years 1790–91 the engineer John Wakefield attempted to curtail the excessive smoke issuing from Britain's first steam powered mule-spinning factory. His endeavours met with a modicum of success, but only at a price. As Wakefield explained to the Select Committee on Steam Engines and Furnaces: 'I turned my attention first to that subject [smoke prevention] with Boulton and Watt's assistance (it was one of their engines), we consumed a part of the smoke, but it took more coals by ten per cent than the old mode.'[22] Wakefield charged that, as abatement was seen to be unprofitable and technically complicated, 'the idea of burning smoke was given up and treated with contempt' by the city's mill owners.[23] However, by the 1840s Manchester's reformers were confident that the latest smoke abatement equipment could produce the 'sugar plum' of fuel economy. Charles Wye Williams, addressing a meeting of MAPS in 1842, reported that under test conditions his method of smoke abatement was very effective in reducing coal bills. But while his plan was economical on fuel, Williams had to admit to the influential gathering that 'he did not get through the same amount of work in the time, and this day the engine did not do its duty; whereas it had before'.[24]

In the words of the old aphorism, 'Time is money'. As coal was relatively cheap and plentiful in the Manchester region, the majority of manufacturers were unwilling to lose valuable time and sacrifice engine performance to make modest savings on their outlay for fuel. Productivity was the unrivalled priority, and not only for the local industrialists. The region's workforce also had a vested interest in whether or not the steam engine 'did its duty' every day. The many thousands of 'hands' who worked in Lancashire's textile mills were generally paid piece rates, related to the individual worker's output, rather than fixed daily wages.[25] Coal and steam provided the workers' daily bread, and, like their employers, few employees would have welcomed a falling off in productivity, even in return for cleaner air. Moreover, there were other important factors that led factory owners and their workers to be cautious concerning the efficacy of smoke abatement measures.

Where smoke prevention was concerned, uneven and untrustworthy technical development did little to advance the reformers' cause. For example, the holes for admitting air into Williams' 'argand furnace' often became blocked with vitrified ash; the links in the 'endless chain' of Juckes' moving grates broke repeatedly; and the 'smoke washing' devices all significantly hindered the generation of steam power.[26] In day-to-day operations most abatement technologies appear to have reduced the efficiency of the engine, and, as well as impairing performance, their adoption could also result in inconvenient stoppages and heavy repair bills. Furthermore, the extravagant claims of inventors with regard to the economy of their devices – some held out the promise of over a 70 per cent saving on coal bills – were all too often found to be an economy of the truth.[27] As one industrialist put it, 'you purchase four pennyworth of economy at the cost of sixpence'.[28] The problems of unreliability and unsatisfactory performance dogged innovation in this technological field.

MAPS also appears to have unrealistically assumed that the new smoke abatement technology was available equally to all manufacturers. However, historical research has shown that most cotton firms in Manchester before mid-century were small, vulnerable businesses, with affluent industrial 'giants' such as the Houldsworths, Ashworths, Gregs and Birleys being the exception rather than the rule.[29] The cost of 'modifications' to polluting boiler furnaces could run into many hundreds of pounds, with little more than the promise of a purer atmosphere by way of a return.[30] While the Lancashire cotton industry was the 'engine of growth' driving Britain's Industrial Revolution, textile manufacturing in the early nineteenth century was an extremely volatile and risky business. Booms and slumps characterised economic growth during the period, and between the years 1836–42 there were no fewer than 241 bankruptcies among Manchester's cotton firms – and by no means all were small concerns.[31] It was said in 'Cottonopolis' that 'three men fall to every one that rises', and it is reasonable to assume that few of Manchester's numerous marginal operators were in a position to be able to absorb the extra expense of new plant even had they wanted to.[32] The small backstreet factory, with badly built flues and chimney, inefficient egg-ended boiler, and a furnace which was 'gorged with coal every few hours, and left to itself between times', remained a familiar sight in Britain's industrial towns at the turn of the twentieth century.[33] But a lack of investment in new technologies, as the above quotation implies, was not the only reason that the reformer's object of clean air remained out of reach. The way in which even the most primitive type of steam technology was actually used also played a crucial role in the creation of the 'smoke nuisance'.

Feeding the furnace
The hand firing of furnaces had always been an 'erratic process', and the development of the mechanical stoker was intended to improve efficiency in this area by largely removing the human factor from the equation. Manufacturers, however, did not take them up with any enthusiasm until the late nineteenth century, when more robust models came onto the market.[34] In Victorian and Edwardian Britain the unreliability of the human stoker attracted almost as much attention from the anti-smoke activists as did the promotion of new technology. While even the best designed and managed furnaces emitted dense black smoke when they first started up, once a steam boiler was running steadily the reformers believed that there was no good reason for it to produce much smoke.[35] It was commonly acknowledged that the stoker, by diligently adhering to the following procedures, could generate steam without producing clouds of sulphurous black smoke:

(1) By spreading small quantities of fresh fuel evenly and often over the surface of the grate.

(2) By ensuring new charges of coals were of a uniform size not larger than a man's fist.

(3) By always keeping a thickness of four inches of bright fuel burning on the grate, and never allowing the fire to burn low or into holes.

(4) By keeping the fire bars free from clinkers so as to admit a free flow of air to facilitate 'efficient combustion'.[36]

But in practice stokers rarely attended to their duties as skilfully or attentively as the reformers would have liked. The daily grind of feeding the furnace was very arduous and uncomfortable work, as a former London stoker vividly recounted to Henry Mayhew in 1861:

It was dhreadful hard work, and as hot, aye, as if you were in the inside of an oven. I don't know how I ever stood it. Be me soul, I don't know how anybody stands it; it's the divil's place of all you ever saw in your life ... the hissin and the bubblin of the wather, and the smoke and the smell – it's fit to melt a man like a rowl of fresh butther. I wasn't a bit too fond of it, at any rate, for it 'ud kill a horse.[37]

Stokers spent no more time in close proximity to the intense heat of the demanding fires than they needed to, and in many cases they had other tasks to perform around the works in addition to their firing duties.[38] That the urban atmosphere was so heavily polluted is a strong indication that most stokers quickly shovelled the coals into the furnace with little regard for 'best practice', and then retreated. Manchester's thickening smoke cloud was in no small measure the result of the stokers' understandable desire to protect themselves from the scorching heat.

From the outset MAPS was aware that where firms had installed newly developed smoke prevention devices, stokers did not necessarily use these technologies in the way that their designers had intended. Henry Houldsworth addressed the following comments to the society's inaugural meeting:

For the last twenty years, scheme had succeeded scheme; and the general impression was, that they had decidedly failed, although he had given testimonials on various occasions, even twenty years ago, of smoke having been consumed; but in practice it was found difficult, if not impossible to secure the attention of servants; and after the eye of the master was removed, the smoke issuing from the chimney assumed the same black hue as before.[39]

In relation to the work practices of the stoker, Thomas Cockshott Rusher, Nuisance Inspector at Leeds, told the Select Committee on Smoke Prevention of 1845 that, 'The masters generally are in their own counting-houses: they do not often interfere'.[40] Perhaps because the boiler room was so very hot and dirty, factory owners were rarely criticised for not supervising their uncooperative workers more closely. Nevertheless, Houldsworth and MAPS were optimistic that the new generation of smoke prevention devices would be more successful in abating smoke and saving fuel.[41] But both reformers and inventors alike had underestimated the way in which the users of smoke consuming technologies would continue to resent and resist their adoption. For example, the stokers who fed the Williams type of apparatus were generally hostile to a technology that kept them at the hot furnace opening and shutting valves to the air inlets.[42] Designers such as Williams paid scant attention to the needs of the sweltering workers and the 'smoke nuisance' persisted.

The failure to provide adequate provision of boiler plant was another major cause of smoke pollution in Victorian Manchester.[43] Despite some improvements in boiler-making to facilitate the use of more powerful engines, such as better riveting, the use of wrought-iron plates, and tubular construction, it was commonplace for factory owners to run boilers that were too small to produce efficiently the amount of steam that their businesses required.[44] By the widely utilised method of 'forcing the fires' a small boiler was made to generate a great quantity of steam and perform the work of a larger one.[45] However, an insufficiency of boiler room resulted in poor fuel economy, a massive loss of heat, and the emission of dense volumes of black smoke, as stokers were forced to continually heap fuel onto the fires in order to keep up a good head of steam, particularly during 'peak periods' of high demand on the engine. Again, in practice steam technology was being used in ways not anticipated by its designers. Indeed, such was the need to quickly build up and maintain a high pressure of steam in these inadequate and overstretched boilers that workmen sometimes fastened down – or even sat on – the machines' safety valves, occasionally with explosive and deadly results.[46]

Generally speaking, the nineteenth-century stoker was badly paid, poorly trained, and recruited from the 'lowest dregs of society'.[47] To the industrialists of the Manchester region, with good access to inexpensive and abundant supplies of fossil fuel from the Lancashire coalfield, the minutiae of fuel economy were not a high priority. And given the unreliability and cost of smoke abatement technology, it is not difficult to understand why many businessmen preferred to employ unskilled workers to service the needs of their inefficient – but highly productive – boiler furnaces. It was in areas isolated from Britain's coalfields, such as Cornwall, where steam engines were used to pump out flood water from tin, lead, and copper mines, that improved fuel efficiency was an urgent priority as high transportation costs made imported Welsh fossil fuel expensive.[48] The Cornish steam engine was widely considered to be the most efficient in the world. In Cornwall furnaces were usually well managed and constructed, and MAPS was informed that their boilers were on average 'four or five times' larger than those used in Lancashire.[49] Cornish steam technology, in comparison with that of the Manchester region, as well as being more economical in terms of fuel consumption also produced less coal smoke. This was not only due to the efficient operation of superior technology, but also because of the mine owners' choice of fuel. Anthracite coal was commonly used to feed the fires of Cornwall's furnaces. This hard coal typically contains around 93 per cent carbon and few impurities, and as a rule it does not produce great volumes of smoke.[50] But in Britain, substantial fields of anthracite were to be found only in south and west Wales and in Scotland. The high cost of transporting anthracite to Manchester would have made this 'smokeless' fuel more than double the price of the ten shillings per ton that local industrialists paid for soft, bituminous Lancashire coal in 1842.[51] Moreover, as supplies of anthracite were limited it could not have provided sufficient power for all of Manchester's booming industries, whatever its cost.[52] Nonetheless, anti-smoke activists often pointed to the efficient use of steam technology in the Cornish mines as evidence that smoke prevention was possible in early Victorian Britain. And it is becoming clear, even allowing for the fact that anthracite was not widely available, that Manchester's intensifying smoke cloud was not altogether an 'inevitable' consequence of the process of raising steam.

The function of tall chimneys

Coal smoke billowing from the massed ranks of Manchester's industrial chimneys was, on the one hand, irrefutable evidence that air pollution was not being effectively tackled at source – in the furnace. On the other, the construction of tall chimneys tacitly acknowledged that the production of smoke was to some extent an unavoidable, if undesirable, consequence of industrial activity. Tall industrial chimneys were in part designed to perform the function of reducing local air pollution by discharging smoke emissions high into the atmosphere. In terms of steam technology being shaped by contemporary environmental values, it was commonly thought that the sulphurous black smoke would then be diluted and dispersed by the prevailing winds. But Lancashire's factory chimneys, as previously discussed in part one, were none too successful in performing this task. Yet this was not their only or even their primary function. The main goal of industrialists in constructing towering factory chimneys, with some standing over 300 feet in height, was to increase the performance of their steam engines.[53] Simply put, as steam technology advanced, manufacturers needed to extract the maximum amount of heat from their fuel to generate more and hotter steam to power bigger engines. Tall industrial chimneys were built in order to increase the draught to the boiler furnaces, and consequently the intensity of fuel combustion. A number of other variables also influenced the height to which any given chimney was raised, including the opportunity for an industrialist to enhance his social status, but the horsepower of the engine involved was usually the most essential part of the equation.[54] However, 'better combustion', where a plant was badly constructed and 'carelessly' operated, was only achieved at a price. For example, Charles Wye Williams advised MAPS that until Manchester's factory owners installed larger boilers, 'much of the heat generated by the improved combustion' would continue to escape up their smokestacks unapplied, especially if the draught from the lofty chimney was too strong.[55]

Household air pollution

By the early 1840s, Manchester had around 500 polluting factory chimneys to which MAPS almost exclusively devoted its energies and attention. The 'wasteful' domestic fire does not feature on MAPS' agenda for change, despite the existence of a proven technology that greatly reduced smoke emissions from the home. It was widely acknowledged that the closed stoves that warmed the homes of townspeople in Europe and America consumed a smaller amount of fuel and produced less in the way of dirt and smoke than the British open fire.[56] So why did the city's reformers place the issue of household air pollution on the back burner? First of all, the thick black smoke vomiting forth from hundreds of industrial chimneys was a much more visible problem. In 1843, for example, Captain A.W. Sleigh described the city's domestic fires thus: 'They do not appear of sufficient magnitude to attract attention … the contrast is very great between the small quantity of smoke coming from private chimneys and the dense clouds of smoke continually issuing from the factory chimneys.' [57] In addition, as both parts one and two make clear, the fireplace was the hub around which domestic life revolved. A strong cultural attachment to the 'cheerful' open coal fire meant any prospective switch to 'smokeless' technologies was likely to prove highly unpopular with Lancashire's householders. However, another key obstacle to bringing about changes in home heating has not yet been considered: the domestic chimney's important role in ventilating the overcrowded dwellings of the poor.

The pre-Pasteurian miasmatic theory of disease held that the putrid stench that pervaded the stagnant air of the enclosed courts and narrow streets of Britain's factory towns caused fever, sickness and death. By creating a bracing circulation of the air in densely populated areas, some contemporaries thought that industrial chimneys were positively beneficial to the health of the nation. And if their actions helped to revitalise the vitiated urban atmosphere out of doors, it was commonly believed that the domestic chimney performed a very similar role inside the Victorian home. In the overcrowded houses of Manchester's working classes, where windows and doors were stopped-up tightly to keep in warmth, as well as to keep out smoke and dirt, fresh air was in short supply. Sanitary reformers claimed that 'close' atmospheric conditions indoors constituted a threat to health every bit as deadly as the 'putrescent effluvia' that rose from sewer-rivers, cesspools and drains.[58] They frequently argued that dangerous 'carbonic acid gas', (carbon dioxide formed during respiration), was insidiously polluting the air of the ill-ventilated houses of the poor.[59] In addition, it was commonly assumed that respired air contained a small proportion of deadly 'organic poison', the putrefying 'animal refuse matter' of the human system.[60] The editor of the *Builder*, George Godwin, neatly summed up the essence of this popular belief in 1859 when he declared, 'Air once breathed is poison'.[61] The carbonic acid gas in respired air was thought to 'act like a narcotic poison, producing drowsiness, which sometimes ends in death'.[62] The symptoms of prolonged exposures to this insalubrious gas were a 'squalid hue' to the skin and 'sunken eyes', accompanied by a general malaise and a predisposition to contract disease.[63] The stale domestic atmosphere was commonly understood to become less and less wholesome with every breath one took:

> By the care we take to shut out the external air from our houses we prevent the escape of the deteriorated air, and condemn ourselves to breathe again and again the same contaminated, unrefreshing atmosphere. Who that has ever felt the refreshing effects of the morning air can wonder at the lassitude and disease that follow continued breathing of the pestiferous atmosphere of crowded or ill-ventilated dwellings?[64]

As late as 1878 the Manchester physician Arthur Ransome thought human respiration 'perhaps the most important of the sources of the pollution of air'.[65] Without a constant circulation and replenishment of the air in the cramped homes of the poor, it was believed that illness and death would be regular visitors. The domestic hearth was recognised to be one of the major means by which a fresh supply of air could enter and revivify the atmosphere of a 'suffocating' room.

In 1843 William Hosking, Professor of Architecture at King's College, London, informed the Royal Commission on the Sanitary State of Large Towns and Populous Districts that, 'Nothing makes a room so sweet and wholesome as an open fire in it.'[66] The draught created by an open coal fire was believed to be of crucial importance to the physical well being of the masses. It drew harmful atmospheric impurities out of a stuffy room and discharged them from the chimney.[67] Large quantities of 'vitiated air' were carried off by the action of the fire, to be continually replaced by the same volume of fresh air drawn into the habitations of the poor through small cracks and crevices that they had neglected to stop up. The traditional open fire was widely considered to be a vital source of household ventilation, particularly in the back-to-back houses and cellar dwellings of Manchester and other factory towns. Indeed, Arthur Ransome went as far as to state that without the benefit of the open fireplace, 'I am pretty certain that half Manchester would die of suffocation in the course of a winter's night.'[68] During the 1840s some contemporaries were already touting the smokeless fuel coke, which could also be burnt in an open grate, as an alternative to bituminous coal. But it was a very unpopular fuel with the local populace as it was difficult to ignite, and it did not make a good blaze in the hearth. Moreover, in early Victorian Britain coke, like its more expensive 'smokeless' counterpart anthracite, was available only in very limited quantities. There were, however, closed stoves on the market that were constructed to burn soft coal, and that produced warmth far more efficiently and with less smoke than the customary British fireplace.[69] Cost put such stoves beyond the modest means of many urban dwellers, but the notion that they were poor ventilators was undoubtedly a major reason why reformers did not strongly endorse them. In 1843, for example, Dr Andrew Ure stressed the dangers to health of the stove-heated rooms of Germany, Russia and Sweden:

> ... their imitation in this country [would be] the greatest of evils. The sallow and withered complexions of the people most subjected to the influence of these stoves, their headaches and dyspeptic ailments are well known to observant British travellers, who perpetually contrast the foul stagnant air respired in these apartments, with the fresh invigorating atmosphere of an English parlour, as heated by the open cheerful grate ... I was led to conclude that [closed stoves] caused an atmosphere incompatible with comfort, health, and longevity.[70]

So powerful were concerns about inadequate ventilation and 'rebreathed air' being harmful to health that they still remained strong in the early years of the twentieth century.[71] Although the primary role of the open hearth was to provide warmth, it was also thought to perform a key 'hygienic' function in ensuring the circulation of fresh air in the overcrowded homes of the working classes. This factor should not be overlooked when one considers why MAPS (and later anti-smoke societies) found it so difficult to excite local and national government interest in promoting alternative 'smokeless' technologies in the home.

The politics of smoke pollution

Determined to reduce urban atmospheric pollution, MAPS' initial strategy was to use its influence to persuade Manchester's industrialists to adopt smoke prevention measures voluntarily. The society's decision to take a conciliatory approach to the problem is unsurprising when the contemporary political climate is taken into account. At the leading edge of technological innovation, the dynamism and enterprise of industrialising Manchester was reflected in its advancing political creed. The fundamental tenets of the distinctive 'Manchester School' of social and economic political theory, already well established by the 1840s, were *laissez faire*, free trade, a commitment to competitive individualism, and an antipathy to the interference of central government in local affairs. The extent to which this set of principles was to become predominant in

Victorian Britain – or even in Manchester – can be exaggerated. However, it is fair to say that the ideas of the city's business community provided the foundations of a national political orthodoxy that endured for much of the nineteenth century.[72] Influential politicians of the 'Manchester School', such as Richard Cobden and John Bright, firmly wedded to the principles of *laissez faire* and free trade, argued that political intervention in commercial and social affairs was undesirable and best avoided. Simply put, their aim was to create a business environment conducive to economic growth by keeping industry free from burdensome rules and regulations. MAPS would have been swimming against the tide of middle class opinion by elevating its threat to 'apply legal means' to curb smoke emissions to the top of its list of goals. Indeed, many of this pressure group's own supporters approved of using the law to combat air pollution only if it could be proven that smoke prevention would benefit Manchester's factory owners financially. The Reverend Richard Parkinson, for example, insisted that, 'Unless they could demonstrate that it was to the interests of the producers of smoke to consume it, it would be an act of tyranny to compel them to do so'.[73] Yet, despite the on-going battle against mechanical breakdowns, lost time, and diminished engine performance, MAPS quickly concluded that the abatement of smoke was both possible and profitable if new prevention technologies were carefully and correctly handled.[74]

The results of the tests undertaken by the engineer Charles Wye Williams had persuaded the society that his method of smoke control was effective in economising coal consumption and bringing down fuel bills. Furthermore, to spur those 'practical men' who were rightly suspicious of the inflated claims of both reformers and inventors into taking action, two of Williams' smoke prevention devices could be seen to be working efficiently at Henry Houldsworth's Manchester factory.[75] Following a public meeting of MAPS at the Royal Victoria Gallery in 1842, an upbeat editorial in the *Manchester Guardian*, renowned for its coverage of business news and issues, argued that existing smoke control technologies could greatly reduce the city's smoke cloud without hindering the operations of its factory owners:

> From what we have seen and learned in reference to some recent inventions, we feel perfectly satisfied, that, even in cases where there is a scanty allowance of boiler room, a very complete combustion of smoke may be obtained, without in the slightest degree diminishing the pressure of steam, or the efficiency of the engines; and parties who persist in poisoning the atmosphere are now left without the slightest excuse for their misconduct.[76]

Soon afterwards MAPS endeavoured to get local businessmen to sign a declaration that expressed their 'willingness to co-operate with the association, and to give [their] best attention to such measures as may be suggested for abating the nuisance'.[77] But only thirty-five manufacturers from the Reverend Molesworth's parish of Rochdale put their names to this voluntary agreement, with just three going on to install the latest smoke prevention equipment in their works.[78] The political philosophy of the 'Manchester School' undoubtedly contributed to this poor response.

The competitive, *laissez faire* ideology of the age meant that MAPS thought it improper to appear to endorse any one specific smoke control device to the detriment of others. The 'argand furnace' developed and patented by Charles Wye Williams, and his work on the theory of 'perfect combustion', did feature prominently in MAPS' campaign against air pollution. But at the same time the reformers pointedly 'disclaimed any connection with or favour towards any particular plan for the prevention of smoke'.[79] For example, when asked by William Mackinnon, Chairman of the Select Committee on Smoke Prevention of 1843, if, in his opinion, coal smoke could be abated, Molesworth, the Chairman of MAPS, replied:

> Yes; I do not conceive that it is a matter of opinion at all, it is a matter of fact. I have seen it remedied by Mr Williams's patent. I do not say wholly; I suppose there will be partial smoking at the time of firing; but three-fourths of the evil is remedied by Mr Williams's patent, also by Mr Hall's, Mr Waddington's, and Mr Rodda's.[80]

The reformers' reluctance to take a stronger lead in recommending effective smoke prevention appliances meant that MAPS largely failed to dispel the confusion and uncertainty that obscured the different paths to cleaner air. Another factor that deterred mill-owners from signing the reformers' voluntary declaration of cooperation was that they feared that it 'might be made the basis of their being coerced, or of their being obliged to adopt one of the plans now in operation'.[81] And following the society's failure to attract large numbers of industrialists voluntarily to the cause of smoke abatement, compulsion was the next step that some campaigners had in mind.

Legal responses to smoke

In an unfavourable socio-political climate the Manchester Association for the Prevention of Smoke, unconvinced that heavily polluted air was a necessary concomitant of industrial 'progress', began to press strongly for decisive legal intervention from the national government. MAPS petitioned the House of Commons to appoint a select committee to inquire into the 'smoke nuisance' in 1843, and the society made it clear that it supported the introduction of fresh legislative measures to tackle the problem.[82] By this date, the existing laws against the excessive production of smoke were plainly failing to protect Manchester's citizens and environment against the harmful effects of industrial pollution. However, while the legislature did not welcome air pollution, the courts were reluctant to encumber modern industry with crippling damage actions. The complex interplay of forces that shaped innovation in steam technology similarly fashioned the course of changes in pollution nuisance law, and Britain's legal system was evolving to facilitate brisk economic growth in its centres of industry.

'A process of compromise': the Common Law
In early nineteenth-century Manchester there were three main avenues open to those actively trying to curb air pollution by legal means: the Common Law, the Court Leet, and the Police Commissioners. With regard to the Common Law of nuisance, a number of detailed legal studies have clearly shown that the courts 'embarked upon a process of compromise' with the Industrial Revolution.[83] Up until the end of the eighteenth century the Common Law courts had accepted that there was a natural right for an individual to enjoy clean air and pure water on their own property. It was no defence for a businessman causing nuisances to claim that his operations were of public utility and benefit.[84] Thereafter, a growing reliance on coal-fuelled steam power in the new factory towns provided the impetus for a shift in the focus of traditional nuisance law. As smoke and other forms of industrial pollution worsened in the manufacturing districts of Britain, the dilemma faced by nuisance law judges was, 'How best to reconcile the often conflicting goals of environmental quality and business growth?'[85] The increasing use of the doctrines of 'prior appropriation of land' and 'social-cost balancing' by early nineteenth century judges weakened the plaintiff's right to protection, and a rigid interpretation of liability was abandoned in industrial areas. The notion of 'prior appropriation' meant that where it could be shown that factories had been established in a neighbourhood for many years, it became extremely difficult for incomers to get compensation for damage to their property or for 'severe personal discomfort'.[86] Indeed, Victorian judges often stated that 'life in factory towns required more forbearance than life else-where'.[87] The utilitarian concept of 'social-cost balancing' allowed the courts 'to weigh the costs of imposing an injunction on a polluter against the benefits of abating the pollution'.[88] Judges making decisions using a balancing approach were well versed in the political beliefs, economic affairs, and cultural values of the age.[89] In Manchester and other factory towns the benefits of abating smoke were thought to be more than outweighed by the possible negative repercussions of pollution injunctions for industrial growth. Thus, traditional restrictions on

economic enterprise that had previously shielded people and the environment from the injurious 'costs' of industrialisation were relaxed. There is little doubt that judicial thinking was attuned to the predominant narrative of wealth, subscribing to the idea that urban dwellers had 'implicitly bargained away a pollution-free environment in return for the benefits of life in modern cities'.[90] The Common Law became ineffectual because the new industrial society had made a pragmatic trade-off: dirty air in return for economic success, jobs, and consumer goods.[91]

As the number of smoking chimneys rapidly multiplied in Britain's industrial towns, relatively few air pollution cases were brought before the Common Law courts. In the ninety-year period after 1770 there were, on average, only one or two actions in England every ten years.[92] Cost was a major factor in dissuading potential complainants from entering the legal lists. Court actions were both time-consuming and very expensive to prosecute. A plaintiff's legal bills could escalate to several thousands of pounds.[93] What is more, as polluting factory chimneys not only mushroomed, but were also constructed to reach new heights, establishing 'cause and effect' in such cases became problematic. By the 1840s, the scientist Robert Angus Smith was beginning to monitor concentrations of harmful pollutants in Manchester's air and water. But his generalised experiments did not constitute the type of rigorous investigation that could conclusively prove a link between a defendant's industrial operations and the damage caused to a plaintiff's property or personal well being.[94] In heavily industrialised regions, where sulphurous smoke constantly issued from a whole host of tall chimneys, it was by no means an easy matter to identify the guilty party, or, as the law demanded, to estimate precisely how much damage a business was responsible for causing.

Sensory perceptions of air pollution were adjudged to be important evidence in nuisance cases. And prosecuting counsel usually produced numerous witnesses in a painstaking attempt to reconstruct the destructive path of smoke and noxious gas emissions in court.[95] Plumes of acidic smoke were observed by witnesses to damage buildings, clothes, and household goods, and to lay waste to great swathes of vegetation in and around the expanding industrial towns of south-east Lancashire. In addition, the smoke fumes that irritated the eyes, skin, nose, and throat of these aggrieved contemporaries were also alleged to seriously impair their health. For example, at the trial of the Manchester alum manufacturer Peter Spence, accused of polluting the air in the 'favourite suburban residence' of Broughton, the court heard that:

The greyish vapour from the defendant's chimney did not rise as high as ordinary smoke. Its smell was pungent and offensive ... [it] rolled along in a compact body, according to the direction of the wind, and which, wherever it struck, produced effects injurious to vegetation; produc[ing] also burning and sore throat, while the smell mixed with that of the hydro-carbons, was intolerable.[96]

However, other witnesses living in close proximity to Spence's works were in turn called by defence counsel to testify that they were in robust health, and that they had experienced little discomfort from the defendant's chimney. Many of the components of the wealth and well-being story line were regularly invoked by those speaking on behalf of industry. The counsel for the defence in the Spence case wasted no time in apprising the court of the fact that 'these foul smells were the workmen's daily bread'.[97] In 'borderline' suburban areas where dense smoke was not the norm, and industry was encroaching upon the amenities of the affluent inhabitants of a town, a judge's decision might still go either way and it could hinge upon the convincing nature of this type of lay, experiential evidence. And while both prosecution and defence counsel generally called upon leading 'scientific experts' to comment on the theoretical causes and effects of atmospheric pollution, their often conflicting testimony was not always fully understood by nuisance judges and in some cases it was completely discounted.[98] In the event of a lawsuit being unsuccessful, costs were normally awarded against the plaintiff.[99] Under such circumstances, only the wealthy could afford to contemplate bringing a nuisance action in the Common Law courts.

A few principled landowners fought hard against the incursions of polluting industry, such as Sir John Gerard, who took the alkali manufacturers of St Helens to law on a number of occasions.[100] But the majority of landowners with estates in the vicinity of the expanding factory towns, rather than 'stand on their rights' and engage in a potentially ruinous battle with Britain's manufacturers, either sold off their property or rented it out, usually at 'greatly enhanced values'.[101] Somewhat paradoxically, the price of land in central Manchester was rising as coal smoke and acid rain degraded its natural and built environment.[102] That businessmen were willing to pay high prices for smoke-blackened property had smoothed the path for the exodus of the landed and moneyed classes away from the polluted environs of Manchester. By the end of the eighteenth century the only remaining family of antiquity residing near to the city were the Traffords at Trafford Hall. Sir Oswald Mosley, Lord of the Manor of Manchester, had retreated to Rolleston Hall in Staffordshire: his only real interest in the city was 'the collection of rents and fines'.[103]

Local responses: the Court Leet and the Police Commissioners
While the obstacle of cost put recourse to the Common Law courts far beyond the reach of the city's lower classes, discontented locals could make their complaints about air pollution known to the Court Leet of the Manor of

Manchester. At the turn of the nineteenth century the jurors of the Lord's Court of Sir Oswald Mosley were active in presenting and fining mill-owners for emitting 'great quantities of smoke and soot' from their factory chimneys. Although the procedure of the Court Leet was cumbersome, it could be activated without great expense as no lawyers were present and no witnesses were heard.[104] A jury, made up of around two dozen of the 'more respectable and substantial residents' of the Manor, presented offenders on the basis of 'their own knowledge', occasionally prompted by information received from fellow townsfolk.[105] In 1801, after viewing a number of industrial concerns, the Court fined or 'amerced' some ten cotton-spinning factories and four other enterprises for smoke offences, mainly in the sum of £100.[106] This, however, was the high-water mark of the Court Leet's activity with regard to penalising Manchester's unregenerate 'smokers'. Between 1802 and its last meeting in 1845 the Lord's Court prosecuted, on average, just one or two smoke nuisance cases yearly, with the fines it imposed becoming progressively lighter over the decades.[107] John McLaren has postulated that a strong sense of civic duty among the Tory mercantile and manufacturing élite that dominated the Court might explain its relatively vigorous behaviour in early nineteenth-century Manchester.[108] Nevertheless, in subsequent years the governing Tory oligarchy did little to check the misuse of steam power in the industrialising city. For, as leading businessmen, often with cotton mills of their own, 'they tended in some measure to share the economic beliefs of their Liberal rivals'.[109] Whether a high church Tory or a nonconformist Liberal, 'All Manchester's leaders owed their wealth to the same economic miracle'.[110] A tolerant, *laissez faire* attitude towards polluting industry cut across party lines. Unsurprisingly, there was no sudden upsurge in smoke nuisance prosecutions after the Tory stranglehold on local political institutions was broken by the free trade Liberals in 1838, when Manchester became a municipal borough. From the outset, the administration of the law was in the hands of large-scale merchants and manufacturers who were unwilling to slow the expansion of trade and industry in 'Cottonopolis' by meting out harsh punishment to firms that polluted the air with smoke.[111]

The business community also controlled Manchester's most important governing body, the Police Commission. Indeed, down to 1818 there was a great deal of 'administrative solidarity' and a substantial degree of overlap in personnel between the Court Leet and the Police Commission.[112] Manchester's Police Acts of 1792 and 1828 empowered the Commissioners to,

> ... take any steps which may be necessary for compelling the owners and occupiers of steam engines and fire engines to construct the fireplaces and chimneys thereof respectively in such a manner as most effectually to destroy and consume the smoke arising therefrom ...[113]

The provisions of the Acts were never vigorously enforced. During the summer of 1800, however, the Commissioners did appoint a large committee 'to attend to and report all Nuisances'.[114] Within one month the Committee on Nuisances had presented the Police Commission with a comprehensive report. It cautioned that in Manchester,

> ... the increase of ... steam engines, as well as the Smoak issuing from Chimnies used over Stoves, Foundaries, Dressers, Dyehouses and Bakehouses, are become a great Nuisance to the town unless so constructed as to burn the Smoak arising from them, which might be done at a moderate expense.[115]

The denunciation of the smoke nuisance contained in the Committee's report was almost certainly the catalyst for the short-lived burst of activity the following year by the Police Commission's close ally the Court Leet. Where the suppression of coal smoke was concerned, down to 1843 when the Police Commission surrendered its powers to the new Borough Council, it was no more inclined to take strong action against the owners of polluting factory chimneys than was the Lord's Court. This is confirmed by Captain A.W. Sleigh, former Assistant Commissioner of Police at Manchester, who in 1843 admitted, 'by the local [Police] Act; there are fines on chimneys that are allowed to smoke beyond a certain density; ... but I never recollect having heard of its being put in force'.[116] When MAPS was formed, the city's main legal avenues to smoke abatement were not well travelled and they led mainly into blind alleys.

The 'best practicable means' of abatement?
MAPS, through its petitioning, testing, and well-publicised meetings, undoubtedly succeeded in raising public awareness of the waste that resulted from the excessive production of coal smoke. So much so that, in advance of national legislation, a new smoke-control clause was included in the Manchester Police Act of July 1844. An anti-smoke clause inserted into local Acts passed earlier in the 1840s at Derby, Leeds and Bradford had provided Manchester's authorities with a suitable template.[117] Following the wording of the clause in these local Acts almost to the letter, section 75 of the Manchester Police Act insisted that 'the best practicable means' for preventing smoke be employed by its industrialists or,

> ... every person so offending shall forfeit and pay the sum of 40 s[hillings] for and in respect of every week during which such furnace or annoyance shall be so used and continued, after one months notice shall have been given to him by the council to remedy or discontinue same ...[118]

However, it was common knowledge that this smoke-control clause had failed utterly in its object of abating air pollution in Derby, Leeds and Bradford long before it was passed at Manchester. The Reverend Molesworth, debating the anti-smoke provisions in the Leeds Improvement Act of 1842, (alongside those of Derby and Bradford), with a member of the Select Committee on Smoke Prevention in July 1843, was asked, 'Are you aware ... the several Acts you have mentioned have not succeeded?' Molesworth responded to the question in the affirmative.[119] At least twelve months before the passage of the Manchester Police Act, the 'best practicable means' clause it contained was known to be just another *cul-de-sac* as far as smoke abatement was concerned.

There were several reasons why this anti-smoke clause was found wanting. To begin with, the terms of the legislation were ambiguous and nonspecific; setting out neither the levels of smoke that constituted an offence against the Act nor the type of smoke prevention technology that ought to be installed. What is more, no scientific apparatus existed to accurately quantify smoke emissions. As J.A. Roebuck, MP for Sheffield, pointed out in 1849, 'They had no gauge for smoke. It was not like heat, which could be measured by a thermometer.' [120] Instead, nuisance inspectors in industrial towns were charged with observing the density of smoke issuing from tall chimneys. In the early 1840s, Manchester's inspectors collected evidence in nuisance cases by 'shading on a form the number of minutes during every hour that a chimney continues to emit smoke'.[121] Smoke emissions were classified as either dense, moderate or none. In order to obtain a conviction under the terms of the local Act, it was,

... essential to prove that a furnace continues to smoke for a whole week, without interruption ... [and] it is almost impossible to get up the evidence of the emission of smoke for a whole week; and the failure of a single day, or even hour's proof, invalidates the evidence ... [122]

Ironically, when air pollution was at its impenetrable worst, particularly during periods of foggy weather, it became impossible to bring prosecutions against the owners of smoky chimneys. For example, the Select Committee of 1843 was informed that an attempt to identify polluters at Manchester was disrupted for five days when thick smoke completely obscured the tall factory chimneys from view.[123] Not infrequently, smoke inspectors were unable to pick out the 'guilty chimneys' of businesses transgressing against the law.

Moreover, so as not to check unduly the acceleration of industrial growth, contemporary understanding of the 'best practicable means' of smoke abatement did not mean 'the best conceivable method, nor the best available method. It meant that method which the manufacturers felt they could install at a cost they believed reasonable'.[124] Complex technical and economic factors were directly involved in the thorny decision making process concerning whether or not an industrialist was actually using the 'best practicable means' to prevent smoke emissions. To compound this problem, the vast majority of smoke inspectors in the employ of local councils were technically incompetent.

Most recruits had little or no relevant engineering, scientific, or business knowledge to draw upon as they began inspecting factory chimneys. On the whole, they were as badly trained and almost as poorly paid as the stokers whose smoke emissions they monitored.[125] As late as 1881 Mr Rook, Superintendent of Manchester City Council's Nuisance Department, readily acknowledged that 'his inspectors as a rule knew nothing of the apparatus with which firemen were dealing'.[126] The fact that 'the inspectors did not turn out to be persons capable of forming an opinion as to the efficiency of the machinery set up to prevent imperfect combustion' led many Victorian magistrates to be dismissive of the evidence that they provided.[127] In contrast, manufacturers who argued in court that they had done their utmost to curtail excessive air pollution from their plant by employing technological measures that were commensurate with their financial means, however rudimentary, were usually assured of a more sympathetic hearing.[128]

Furthermore, two other major weaknesses helped to render the anti-smoke clause in these local Acts a 'dead letter'. The first was the trifling nature of the penalties imposed for polluting the air where a case was proven. Unlike the £100 fines handed out by the Court Leet at the turn of the nineteenth century, the new 40-shilling (£2) penalties were widely considered to be 'so small as to be practically useless'.[129] In addition, the stipulation that an offender had to be given one month's notice in writing before every prosecution meant that no more than six fines could be imposed annually by the magistrates, amounting to the trifling sum of £12 in total. Sir Henry De la Beche and Dr Lyon Playfair's 1846 report on Smoke Prohibition in large towns argued that a £12 maximum penalty was,

> ... a sum so small, that a refractory smoke-maker would be willing to pay it rather than be put to the greater expense of altering his boilers or furnaces. The smallness of the fines acts prejudicially to the success of the Acts in Derby, Leeds, Manchester, &c. ... [and] when we visited [Leeds] in the present year, the chimnies were pouring out black and opaque smoke, as if no Act prohibiting it existed.[130]

A second reason why the local Acts remained largely inoperative, as De la Beche and Playfair also emphasised in their report, was that in the industrialising north the local magistrates were seen as an integral part of the smoke problem, rather than as key figures in its solution:

> ... [with] reference to the local magistrates ... when it is considered that these magistrates themselves are frequently the principal smoke-makers, and that they, from habit, are so accustomed to the smoke of factories as to forget the extent or magnitude of the nuisance, it is not surprising, however honest may be their intentions, that the local Acts have not been attended with the results which were anticipated.[131]

In Manchester and many other industrial towns, businessmen were increasingly prominent on the bench as well as pulling the strings in local politics.[132] A brief outline of the interests of one of Manchester's leading magistrates, Hugh Hornby Birley, may serve to illustrate this point. The Tory cotton master Birley was not only one of the bulwarks of local government, but also the first president of the city's Chamber of Commerce. Embodying the entrepreneurial ethos of the Chamber, Birley harboured a 'strong aversion' to the regulation of industry and was closely associated with the distinctive economic creed that the city's Liberals were trying to make their own.[133] In the early 1840s, the family firm of Birleys and Company was one of Manchester's most substantial employers. Birleys was also one of its smokiest factories.[134] Many of industrial Lancashire's magistrates belonged to, or were connected with, the business élite who controlled the local institutional levers of power, and who were most active in propagating the narrative that linked coal smoke with continued industrial growth and economic success.[135] To sum up, a sympathetic local magistracy, obdurate local authorities, laws that were worded ambiguously, the difficulty of proving the source of air pollution and the damaging effects it had on the people's health, and, not least of all, the inconsequential penalties imposed upon offenders seriously weakened these early Acts. And, for that matter, all subsequent Victorian statutory regulation concerning the 'smoke nuisance' right down to the turn of the twentieth century.

'Industry in danger'

Aware that there had been at best a negligible improvement in the atmospheric conditions at Derby, Leeds, and Bradford following the passage of their local Acts, some members of MAPS looked to the state to control smoke pollution from factory chimneys. The Reverend Molesworth, in defiance of Manchester's prevailing *laissez faire* values and strong commitment to self-government, went as far as to suggest that the appointment of smoke inspectors should be taken out of local hands and placed under the jurisdiction of the central government's Factory Inspectors.[136] A science and technology led approach to smoke abatement undoubtedly demanded the selection of independent, technically competent officers to inspect boiler-furnaces and advise sceptical manufacturers on how best to comply with the provisions of any new act. Nonetheless, Molesworth's somewhat heretical proposal came to nothing. Despite two reports from Select Committees on Smoke Prevention in 1843 and 1845 supporting MAPS' position that abatement was technically possible at a moderate expense to mill owners, the reformers were unsuccessful in attaining their legal objectives. Between the years 1844–50 no fewer than six attempts to bring in national anti-smoke legislation were to founder.[137] The well organised manufacturing lobby in the House of Commons, with Manchester's radical free trader John Bright

– a cotton master himself – at its head, succeeded in 'scaring the government' into dropping consecutive smoke control Bills by pressing into service the distress call 'industry in danger'.[138]

MAPS' anti-smoke campaign could hardly have been launched at a more inopportune time, getting underway as it did in 1842 at the lowest point of a deep trade depression. On the whole, industrial recession and extensive unemployment marked the 'Hungry Forties'.[139] While anti–smoke campaigners emphasised the myriad costs of the 'Black Smoke Tax' to townspeople, the pro-smoke lobby in Parliament argued more persuasively that coercing businesses into adopting uncertain and expensive abatement technology at this time could be potentially catastrophic for British industry. George Muntz, master manufacturer and MP for Birmingham, reflected the majority view in the Commons when he warned against pressing smoke producers into altering their plant 'in the present state of trade, when every shilling is an object to the manufacturers, competing as they are with the Continent'.[140] In cotton Lancashire contemporaries were becoming more aware of a growing challenge from 'powerful nations ... whose eagerness of competition is stimulated by view of the rich prizes which we have already won'.[141] Before 1850, however, Britain had no serious rivals for international markets in manufactured goods.[142] Becalmed in the economic doldrums, big businessmen clearly exaggerated early 'concerns' about foreign competition in order to strengthen their defence against constraining regulation. One notable feature of the manufacturing lobby's rhetoric was the way in which it used negative imagery of smokeless chimneys, claiming that 'they would have to give up business altogether if they were not allowed to smoke'.[143]

Through their vocal opposition lobbyists such as Birmingham's Muntz, Sheffield's Roebuck, and J.L. Ricardo, MP for Stoke-on-Trent, succeeded in obtaining exemptions from prospective anti-smoke legislation for the metal, brick, glass, pottery and other trades, because furnaces were used intermittently and/or enormous heating power was required during the manufacturing process. This only served as a spur for Bright to redouble his uncompromising efforts to put an end to talk of smoke control. Successive Bills, he argued, seemed to be aimed at 'nobody but the [textile] manufacturers of the north of England'.[144] Bright contended that Manchester and the textile districts were being unfairly discriminated against and that any abatement act would 'expos[e] the manufacturers to great difficulties and annoyances'.[145] Although the manufacturing lobby often grossly overstated its case, the government was reluctant to hamper the slow recovery of the nation's important wealth-producing industries. It did not wish to jeopardise in any way the pre-eminent position that the British economy enjoyed in the wider world, particularly as there was no hard evidence that conclusively proved a link between smoke pollution and ill health in industrial towns: a 'fact' that Bright brought purposely to the attention of the cautious members of the House of Commons.[146] Under sustained fire from the influential Bright and other opponents of smoke control, a government broadly in favour of the 'Manchester School's' political philosophy repeatedly ran up the white flag with regard to decisive legislative intervention. Moreover, before mid-century there was no major public outcry at this serial failure to bring in effective national smoke abatement legislation.

For local or national government to pass, and, more importantly, rigorously enforce tough laws against air pollution, the reformers needed to harness the weight of public opinion to their cause. Despite favourable press coverage, no such groundswell of popular support for smoke abatement materialised in early Victorian Britain's industrial communities. To most contemporaries industrial smoke represented wealth creation and 'good times', and only to a subordinate degree was it viewed as a symbol of waste. In the lean years of the 'Hungry Forties', Lancashire's working classes shared bitter experiences of clean air as a luxury that they could ill afford. When, for example, many of Bolton's cotton factories closed in the slump of 1841–42, unemployment rapidly spread out from the mills to affect all trades in the town. While some 60 per cent of the town's textile workers were out of work, those hardest-hit were Bolton's bricklayers, with no fewer than 87 per cent made unemployed.[147] The dialect author Edwin Waugh, in a barbed comment directed at those who 'grumble[d] about th' air bein' smooky i' Lancashire', neatly summarised the workers' predicament: 'every factory chimbley wur clear an' cowd ... Ay, – th' air wur clear enough just then, – an' a deeol o' folk had very little else to live on.' [148] Like their employers, the workforce of the Manchester region was highly suspicious of anti-smoke initiatives that might endanger future industrial growth and employment prospects. In the 1840s, then, there was little grassroots support for the authorities to put a stop to the 'smoke nuisance'.

MAPS in brief

After a short-lived burst of activity, the Manchester Association for the Prevention of Smoke faded from public view at around the same time as the Manchester Police Act, with its weak smoke-control clause, was passed in 1844. It is not unusual for a pressure group to lose support and impetus when a goal is attained, and obtaining legislation to curb smoke emissions had become an important objective. However, it is not clear if the society's members considered the passage of legislation that they knew to be seriously flawed a triumph. Probably not, as Molesworth's suggestion that smoke control be taken out of local hands indicates. Nevertheless, that Manchester's authorities took legal action at all shows that MAPS did enjoy some success in getting its message across. In addition, the society's role in testing new abatement technologies and techniques, undertaken in alliance with experts such as Charles Wye Williams, had demonstrated that it was possible to reduce smoke from boiler furnaces and perhaps save a little money on fuel bills into the bargain. Yet many of MAPS' aims were not achieved. No lasting links were developed with smoke abatement societies in other towns, no permanent exhibition showing effective smoke prevention equipment was established, and no dependable statistics were produced concerning the fuel efficiency of the numerous appliances on the market. And given that early smoke abatement devices were often unreliable and expensive to install and operate, MAPS did not manage to dispel the manufacturers' general impression that they were all uneconomic failures.[149]

Ironically, as the science and technology of smoke prevention was becoming better understood the laws prohibiting air pollution were relaxed. This was regrettable, as it is possible that the systematic use of stiff fines might have acted as a stimulus to innovation in the field of smoke abatement technology. However, it must be noted that many anti-smoke campaigners had no desire to 'tax' industry with heavy fines. Manchester's Henry Houldsworth, for example, was among those who supported the introduction of modest £2 fines for smoke offences. Indeed, William Mackinnon, a leading anti-smoke activist and Chairman of the Select Committee on Smoke Prevention of 1843, cautioned Molesworth, who appeared to favour tougher fines:

> ... if we are to legislate to obviate the nuisance, which I think it is the desire of the Legislature to do, ... we may not legislate to bring a very heavy and onerous tax or charge upon the manufacturer which would be injurious to his interests ... You are aware the great point is to do what is most beneficial to the public ...[150]

The law sheltered industry, for fear that the 'suppression [of smoke] might materially injure important branches of our national industry'.[151] Under Manchester's new Police Act, just thirteen cotton firms, including Birleys and Company, were prosecuted and fined the insignificant sum of £2 for making dense smoke in 1845.[152] No legislative action was contemplated against coal smoke from the traditional open fireplace at this time, not least of all because the closed stove, the main technical solution to the domestic smoke problem, was commonly believed to be harmful to the public's health. No wonder, then, that at mid-century a General Board of Health inquiry into air pollution found that 'Manchester is still a very smoky place'.[153]

The legal compromises of the period protected the phenomenal growth of industry in the nation's manufacturing towns, rather than the urban population and environment. The increasing use of the doctrine of 'prior appropriation' produced a 'zoning effect' that saw polluting industries becoming more concentrated in working-class districts, such as Ancoats in Manchester. By 'sacrificing' such areas to boost economic growth, many contemporaries hoped that Britain's leafy middle-class suburbs and picturesque rural landscapes might remain unblemished by smoke and soot. Robert Angus Smith, the government's first Alkali Inspector, argued that while acidic smoke fumes remained a severe environmental problem, 'it may be expedient to adopt the view of abandoning a certain district to them'.[154] H.H. Collins made the case for such a trade-off even more forcefully:

> If the land only is affected, if health is not materially interfered with, if manufactures can flourish with due regard to human health, we had better interfere as little as we can with manufacturers, and let them have as much scope as possible. Seeing how other countries tread on our toes, we must allow a certain quantity of land to deteriorate in [aesthetic] value, whilst hundreds and thousands of labourers are employed earning livelihoods for themselves and their families.[155]

Following Franz-Josef Brüggemeier's approach to industrialisation and its destructive effects in Germany's *Ruhrgebiet*, it is useful to think of the 'protected' Manchester region as an 'industrial reserve' that contained big business, the polar opposite of later wilderness reserves and national parks.[156] Before mid-century both local and national government could have done more to protect the townspeople and environment of industrialising Lancashire from the 'smoke demon'. But, as was the case in the *Ruhrgebiet*, 'Rather than reducing pollution, society adapted to the new industrial conditions'.[157]

The search for solutions continues

By the early 1850s the 'workshop of the world' was back in full production as markets revived. As a result, most of cotton Lancashire's inhabitants enjoyed a rise in living standards after the privations of the 'Hungry Forties'.[158] As smoky prosperity returned to the factory towns, Absolom Watkin, a Manchester cotton merchant, noted in his diary: 'This is the last day of Whitsun week, and the people of Manchester have never enjoyed it more, nor have I seen clearer evidence of general well-being. Our country is, no doubt, in a most happy and prosperous state.'[159] But as business improved, there was still a good deal of laxity in the administration of new laws to combat smoke pollution. Carlos Flick has calculated that from the 1850s right through to 1865 the local authorities at Derby, Leeds, Leicester, Manchester, and Newcastle-upon-Tyne managed to secure on average only four or fewer smoke convictions per year. Even then the fines that they imposed were negligible. In the industrial towns of Huddersfield, Stoke-on-Trent, Sunderland and Wolverhampton no prosecutions were reported for smoke offences between 1850 and 1865.[160] During the mid-Victorian period, as discussed in part one, the smoke cloud was deepening over

Britain's manufacturing centres and gradually extending its reach to begrime the surrounding countryside. At Manchester coal consumption by industrialists and householders had more than trebled in the space of 40 years, increasing from 900,000 tons a year in 1836 to around 3 million tons per annum in 1876.[161] Yet despite the demise of MAPS, the cause of smoke abatement did not 'fall to the ground'. The Manchester and Salford Sanitary Association was to continue the struggle to raise awareness of the dangers associated with air pollution.

The Manchester and Salford Sanitary Association, established in 1852, sought to reduce the twin cities' appalling mortality rates. As the 'smoke nuisance' was thought to be a significant cause of ill health and 'dirty habits', the Sanitary Association joined the fight against it.[162] Smoke and soot was not an inconsiderable obstacle to its primary aim, which was to 'promote attention to personal and domestic cleanliness, to temperance, and to the laws of health generally' among the working classes.[163] Like MAPS, the Sanitary Association was quick to stress the 'pecuniary advantages' which would accrue to factory owners if they would only consume their smoke.[164] However, due to a lack of funds and the more pressing need to induce Manchester's authorities to provide sewers, drains, clean water and an efficient refuse removal service, its anti-smoke campaign was neither systematically pursued nor even its foremost concern. It was an off-shoot of the Manchester and Salford Sanitary Association that gave finding solutions to the cities' air pollution problems a much higher priority.

The Manchester and Salford Noxious Vapours Abatement Association[165]

On 2 November 1876 concerned members of the Sanitary Association and other 'public-spirited' Mancunians met together in the Town Hall to discuss the problems caused by noxious vapours from alkali and other chemical works. The object of the meeting was 'to form in Manchester a branch society to co-operate with the Lancashire and Cheshire – the Liverpool – and other similar associations – for controlling the escape of noxious vapours ... from manufactories, especially by rousing public attention to the evils in question'.[166] The Royal Commission appointed to investigate air pollution from chemical works, which started work in July 1876, was undoubtedly the catalyst for the launch of the Manchester and Salford Noxious Vapours Abatement Association (hereafter NVAA) at this particular time. The focus of the newly formed pressure group soon began to change, however, as the main source of Manchester's problems regarding air pollution was recognised to be coal smoke. In 1876 there were just seven chemical works under the jurisdiction of the Alkali Inspectorate in Manchester and its immediate environs. And even after the passing of the Alkali &c Works Regulation Act in August 1881, which extended fixed emission standards from works that produced hydrochloric acid to include those producing sulphuric and nitric acid, this number had only increased to 32 by 1884.[167] In marked contrast, by the last decades of the nineteenth century there were almost 2,000 industrial chimneys polluting the air of Manchester and Salford with sulphurous black smoke, not forgetting the voluminous emissions from countless domestic chimneys. Less than six months after its foundation, a special meeting of the NVAA resolved that 'the suppression of the Smoke Nuisance' be included in the objects of the society.[168] And from 1881 onwards, the NVAA's energies were 'mainly directed towards the abatement of the most common cause of air pollution, viz., ordinary coal smoke'.[169]

Following the lead of the Sanitary Association, the NVAA adopted a conciliatory, educational approach to its chosen task. Indeed, its strategy closely mirrors that of the Manchester Association for the Prevention of Smoke. Even so, the new society appears to have been largely unaware of the activities of the city's earlier anti-smoke pressure group. The NVAA planned to attack the 'smoke nuisance' from three different directions:

(1) By promoting the development and adoption of efficient and reliable smoke prevention appliances, and by encouraging Manchester's manufacturers to employ properly trained stokers.

(2) By campaigning for more effective laws and regulations to abate coal smoke.

(3) By attempting to educate, inform, and stimulate an active public opinion against air pollution, highlighting in particular the direct and indirect loss and damage that accompanied smoke.

What is more, the NVAA maintained cooperative links with other anti-smoke groups formed in the industrial towns of late-Victorian Britain. But was this new society to surpass MAPS and achieve the object of its reform agenda, clean air?

The technological response to smoke

By the 1880s long-standing concerns about the economic, health, and environmental costs of smoke pollution had been augmented by an apprehension about how long Britain's coal supplies would last and alarm about the 'deterioration of the race'. Coal smoke was roundly condemned by anti-pollution activists as 'barbarous and unscientific' and increasingly portrayed as 'heat and power in the wrong place'. Echoing MAPS, Manchester's reformers were again optimistic that science and technology would be able to provide the answers to the wideranging problems posed by the 'smoke nuisance'. The smoke abatement appliances available to industrialists and householders, although mainly based on the same principles as those in use forty years earlier, were becoming more robust and reliable. Mechanical stokers, for example, were manufactured using stronger materials and they were simpler to operate as they utilised fewer moving parts. Components were also made more accessible to facilitate maintenance and repairs. Moreover, mechanical stokers featured other notable design innovations, such as the incorporation of a mill inside the hopper to crush the fuel to a uniform size, thereby aiding 'perfect combustion'.[170] On the domestic front, closed stoves were now being designed with their own built-in ventilation systems to remove 'foul air' from crowded rooms, while a large selection of modern kitchen ranges offered a 'very small consumption of fuel with rapidity and efficiency of cooking powers'.[171] Inventors and engineers, however, were putting more and more smoke prevention appliances on to the market, in some instances appliances of dubious quality. There is ample evidence to show that more than a few late-Victorian businessmen were extremely dissatisfied with the often costly abatement technology that they had installed.[172] This was not simply because they failed to prevent air pollution (used correctly many devices actually reduced smoke emissions quite considerably), but because the substantial savings on fuel bills that were routinely promised frequently failed to materialise. As in the 1840s, uncertainty and expense were important factors in deterring large numbers of businessmen and householders from using smoke abatement appliances. The NVAA's first major undertaking, the Manchester Smoke Abatement Exhibition of 1882, aimed to demonstrate to sceptical contemporaries that preventing air pollution was uncomplicated and economical and, where industrial equipment was concerned, that it would not interfere with production-oriented goals.

The Manchester Smoke Abatement Exhibition
In December 1881 the NVAA successfully entered into negotiations with the London-based committee that became the National Smoke Abatement Institution to transfer its popular Smoke Abatement Exhibition to Manchester.[173] The show opened at South Kensington, London, on 30 November 1881, and by the time it closed in February 1882 it had attracted no fewer than 116,000 people through its doors, including the Prince of Wales and the Empress Eugénie.[174] As London's air pollution problem was largely the consequence of smoke issuing from household chimneys, the Exhibition understandably focused on promoting domestic smoke abatement apparatus. At Manchester, where factory smoke was still the main cause for concern, the accent was placed firmly upon boosting the technical remedies for industrial air pollution. The Exhibition, which reopened at Manchester's Campfield Market, Deansgate, on 17 March 1882, drew in some 32,000 people before it closed six weeks later.[175] The NVAA, like its predecessor MAPS, was reluctant to 'express a preference' concerning which of the many smoke control devices on offer should be adopted.[176] Instead, in

keeping with the spirit of the age, exhibitors were urged to enter their smoke abatement apparatus in a series of competitive trials, with the winners being awarded prize medals and certificates of merit. Fuel efficiency, pressure of steam raised per pound of coal consumed, and the duration and shade of any coal smoke produced were but three of the many factors the judging panels at London and Manchester took into account. Winners included Edward Bennis's mechanical stoker, which 'greatly increased the production of steam', E.H. Shorland for the smoke-reducing 'Manchester' closed stove, and the Eagle Range Company for its energy saving coal-fired kitchener. The results of these competitive tests were later tabulated and published by the reformers as a serviceable guide for interested parties.[177] But equally importantly, the Exhibitions were devised as a kind of 'public laboratory' where visitors could see these improved technologies in action, working both effectively and economically.

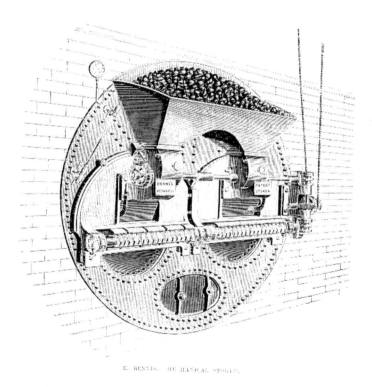

Illustration 10. Bennis's Mechanical Stoker.

Illustration 11. Shorland's 'Manchester' Stove.

THE EAGLE RANGE.

IMPORTANT
ADVANTAGES.

1. Requires no brickwork
flue.

2. Can be used with either
open or close fire.

3. Size of fire can be increased or diminished
as required.

4. Roasts perfectly in front.

IMPORTANT
ADVANTAGES.

5. Oven can be heated
equally in all parts,
or an excess of heat
turned on top or
bottom.

6. Very small consump-
tion of fuel with ra-
pidity and efficiency of
cooking powers.

Highest Awards, SILVER MEDAL and SPECIAL LADIES' PRIZE, 25 GUINEAS, International Smoke
Abatement Exhibition, 1882, and 12 First Prizes wherever shown in Competition.

THE EAGLE RANGE & FOUNDRY CO.

168 FLEET STREET, LONDON, E.C.

Illustration 12. The Eagle Range.

Source for illustrations 10–12: *Report of the Smoke Abatement Committee 1882*, Smith,
Elder, & Co., London, 1883.

Neither the London nor the Manchester Exhibition, however, managed to induce large numbers of Britain's householders or industrialists to adopt smoke abatement technologies. The working-class attendance at the Manchester Exhibition was good, perhaps making up as many as 20,000 of the 32,000 visitors.[178] But the vast majority almost certainly came to hear the band that played every Saturday afternoon and to view such novelties as Tyndall's Musical Flames, rather than out of any real enthusiasm for smoke abatement and the 'cheerless' alternatives to the open hearth.[179] In addition, most businessmen remained unconvinced that an appliance that worked efficiently and produced more steam under 'artificial' test conditions would prove as successful in everyday operations. They argued that the trials undertaken at both Exhibitions, 'conducted under conditions arranged by the patentees and makers', were not a 'safe guide' as to how they would perform 'in actual work'.[180] The bulk of Manchester's factory owners, its numerous marginal operators included, were not swayed by the reformers' claims that 'it would be no loss to them to consume their own smoke, [and] that it was to their interest to do so'.[181] Furthermore, George Davis, District Inspector of Alkali Works and a founder of the Society of Chemical Industry, was singularly unimpressed by some of the devices demonstrated at the Manchester Exhibition. He informed the Society of Chemical Industry that at times 'there was more smoke emanating from the smokeless furnaces and fireplaces at the Campfield Smoke Abatement Exhibition' than from the chimneys of the surrounding area.[182] That the Manchester Exhibition and its statistical report had little long-term effect in combating Lancashire's 'smoke nuisance' is evidenced by the fact that seven years later the NVAA embarked on another major campaign to promote the technology of smoke abatement.

The Committee for Testing Smoke Preventing Appliances
Aware that the technical information published in the report of the Smoke Abatement Exhibitions of 1881 and 1882 was considered to be flawed and unreliable by many of Britain's manufacturers, the NVAA decided to carry out new trials on appliances 'in actual use' at the workplace. To this end the society organised a public meeting at Manchester Town Hall on 8 November 1889, the result of which was the formation of the Committee for Testing Smoke Preventing Appliances. Among the distinguished ranks of the Committee were the Duke of Westminster, Lord Egerton of Tatton, Sir Douglas Galton, President of the Sanitary Institute of Great Britain, and Ernest Hart, Chairman of the National Smoke Abatement Institution. Public meetings in support of its work were held at London, Birmingham, Leeds, Glasgow, Newcastle-upon-

Tyne, and Sheffield, and affiliated committees were also established in the latter four cities.[183] The main object of the new Committee was to arrange for,

> Examinations or tests to be conducted by experts, appointed by the Committee, at places where the appliances for, or methods of, consuming coal smokelessly are already at work, with the object of ascertaining whether under ordinary working conditions – i.e., with ordinary workmen, ordinary coal, and under varying demands for work – these methods or appliances do produce or are accompanied by – (a) Practical freedom from smoke. (b) Reasonable amount of duty. (c) Economy of fuel. (d) Moderate cost in wear and tear and simplicity of construction. (e) Moderate cost of application.[184]

The engineer Professor A.B. Kennedy and Alfred Fletcher, the Chief Inspector of Alkali Works, were among the eminent authorities on fuel combustion appointed to undertake the investigation. The Committee's experts carried out over seventy exhaustive tests of smoke prevention appliances, almost all in operation at industrial premises, and its comprehensive report on the performance and reliability of abatement technologies took over five years to compile – mainly due to a lack of public support.[185]

Although this time around the NVAA had dispensed with the idea of a formal exhibition, numerous 'public demonstrations' of the technologies in everyday use were organised. The latest and most efficacious mechanical stokers designed by Vicars and Sinclair were viewed at work *in situ*, along with examples of cheaper technologies that introduced jets of air into the furnace, and, to show dependability, an original John Juckes stoker that had been in use for over twenty years.[186] During 1889 a key member of the Committee, Herbert Fletcher, demonstrated the mechanical stokers employed at his works in Bolton to nearly 3,000 people, including manufacturers, local health and nuisance committees, magistrates, medical officers of health, and nuisance inspectors.[187] Fletcher maintained that they enabled him to 'get more work' out of his boilers and 'Though there were repairs required, this was compensated for by less labour'.[188] He also asserted that the machines had paid for themselves in just three years, due to substantial savings made on fuel bills.[189] But despite the unstinting provision of 'optical evidence ... of successful smoke prevention' and the stamp of approval of an influential committee of the 'highest scientific authority' many manufacturers remained far from convinced that abatement could be made into 'a paying business'.[190] Late-Victorian methods of abating air pollution were fundamentally unchanged from those of the 1840s, and anxieties over outlay, breakdowns and loss of productivity persisted. For example, one man of letters, writing under the pseudonym 'Mechanical Engineer', informed the *Manchester Guardian* that:

... very many mechanical stokers are thrown out of use after a time and the old method of hand-firing is again restored, the general remark being that, though pleasing and satisfactory at first, the main result in the long run is that the machines are troublesome, wasteful and costly ... it is reasonable that manufacturers should be reluctant to spend £100 per boiler on machines that are exposed to such a high temperature that the depreciation must be enormous.[191]

Similarly, 'Smoke Creator', in contributing to the *Guardian's* ongoing discussion, did not challenge the fact that smoke was substantially reduced at Fletcher's works. However, he did protest strongly that the amount of economy and efficiency Fletcher was claiming for his mechanical stokers went against 'the dictates of common sense and practical experience'. 'Smoke Creator' concluded by offering the reformers some blunt advice: 'By all means let us further the cause of smoke prevention; but this will be more effectually done by refraining from statements that transcend the bounds of credibility.'[192]

Nor did the business community as a whole actively support the Committee's endeavours. The reformers had hoped to raise the sum of £5,000 to make short work of carrying out the tests and they looked to industry for constructive cooperation and cash donations.[193] The anticipated assistance from Britain's manufacturers was not forthcoming. In the end, the impecunious anti-smoke reformers were hard-pressed to bring their self-imposed task to fruition. When the Committee for Testing Smoke Preventing Appliances' final report was eventually published in 1896, *The Times* commented, 'The committee have been disappointed at the smallness of the support they received from steam users, in whose interests the work was undertaken.'[194] This disappointment was compounded when, despite extensive national and local press coverage of its publication, British businessmen continued to show little interest in the Committee's research and only a few copies of the detailed report were sold.[195]

The words that concluded the Committee for Testing Smoke Preventing Appliances' report confirmed – and echoed – the findings of almost every investigation into the 'smoke nuisance' undertaken during the nineteenth century:

... in the great majority of cases the black smoke thrown into the air during the combustion of coal is preventable, either by hand or mechanical firing, and without great cost to the consumer. Often the prevention of smoke is accompanied with a saving of expense ... a manufacturing district may be free from manufacturing smoke – at least from the steam boilers, with which alone the committee have concerned themselves; and ... the suppression of the smoke nuisance means an increased pleasure in life, and would unquestionably add to the health and wealth of the community.[196]

A member of MAPS would not have found anything out of the ordinary in the Committee's summary of its findings. In the conclusion to the Sheffield sub-committee's report on smoke prevention appliances another enduring difficulty surfaced prominently:

> Whilst it is certain that smoke may be almost entirely and completely prevented from steam boiler chimneys, the conditions of working are so varied that no single arrangement can be expected to meet every individual case, and further, whatever device is applied to a boiler to prevent smoke, its success will in a great measure depend upon the intelligent handling and management which it receives on the part of those to whose care it is trusted.[197]

And where the 'intelligent handling' of a boiler furnace was concerned the NVAA was attempting to break new ground.

A national Institute of Enginemen, Boilermen, and Firemen
Careful firing could not prevent smoke emissions from numerous businesses, both large concerns and small, 'where boilers [we]re overtasked'. But in cases where sufficient boiler room was provided, the prevention of smoke could be accomplished 'without any special appliance' if furnaces were skilfully tended.[198] Whether or not a well set up boiler furnace produced volumes of dense black smoke still 'very much depend[ed] on the fireman'.[199] That this backbreaking task was as poorly paid as ever in the late nineteenth century, and still involved prolonged exposure to intense heat, undoubtedly contributed to the sluggish pace of smoke abatement. In 1898 a Lancashire smoke inspector commented, 'These men [stokers] are sometimes very hard worked, and in such cases take but little interest in the prevention of smoke.' [200] However, even at works where manufacturers had installed the latest in labour-saving mechanical stokers, the reformers argued that 'good tools' were being placed in the hands of bad workmen. Firemen, allegedly 'the most conservative class of men living', were continually failing to use such appliances properly. For example, they often overloaded mechanical stokers, thereby wasting fuel and causing smoke. As a result, the nation's industrialists had come 'to believe that they are useless'.[201] Indeed, the NVAA was informed that stokers at Manchester were supplementing their meagre wages by deliberately burning more coal than was necessary in raising steam. '[I]t was a common practice', the reformers heard, 'for coal merchants to pay a commission of 1d or 2d per ton of coal consumed to stokers'.[202] The NVAA soon turned its attention to remedying this ongoing and

perplexing situation. After a fruitless attempt in the early 1880s to interest the Manchester City Council in awarding 'medals or other rewards' to 'careful' firemen, the reformers devised a much more ambitious scheme to tackle the problem.[203]

Acting on a suggestion made by the Glasgow and West of Scotland Smoke Abatement Association, in 1894 the NVAA unveiled its plans to establish a national institution that would organise the 'technical training ... of all workmen entrusted with the charge of boilers, furnaces and engines'.[204] Stokers would receive 'scientific training in the principles of combustion' and the status of the job would be improved by the 'certification and registration' of this class of workmen. Better boiler performance and cleaner air would ensue, as one contemporary put it, when,

> ... brains are put into the handling of the coal shovels; it pays to educate firemen in the most economical methods, and to encourage them by paying adequate salaries ... Firemen have too long been classed among unskilled labourers. The efficient performance of their duties really calls for high skill, and their work rises to the dignity of a profession, – even of an art.[205]

The NVAA was confident that the proposed Institute of Enginemen, Boilermen, and Firemen, which also aspired to act as an employment agency, would secure for employers and the public numerous advantages. These included less smoke, the more economical use of coal, a guarantee of the efficiency of registered workmen and the facilities for obtaining trained men when needed.[206]

Nevertheless, insufficient numbers of employers came forward to support the scheme and its launch was unsuccessful.[207] At Manchester the Board of Directors of the Chamber of Commerce opposed the training of stokers on the grounds that,

> ... there is no general demand. The proposal to create a certificated class of 'skilled workmen' ... is considered objectionable in itself, and would, it is believed, prove very inconvenient in practice. In its immediate effect it would bear hardly upon steam users, who would be required to pay higher wages without necessarily receiving more efficient service.[208]

Not only did the directors of the Manchester Chamber of Commerce object to the idea of paying higher wages to workmen for performing the 'simple' task of shovelling coal. They also feared that a scheme such as the planned Institute, by establishing a body of skilled, organised men, would tend to 'create difficulties in the management of works':[209] not surprising at a time when militant trades unionism was making great strides in Britain, with membership growing from 750,000 in 1888 to over 2 million workers in 1900.[210] This was accompanied by the disquieting, although less rapid, growth of socialist organisations such as the Social Democratic Federation and the *Clarion* societies. The NVAA's design for a national Institute of Enginemen, Boilermen, and Firemen came to nothing due to the opposition of wary employers, who for the most part continued to consume cheap and plentiful coal supplies in a profligate manner. The notion of training stokers was half-heartedly revived in 1911 when voluntary classes for firemen were started up locally at the Manchester College of Technology.[211] However, the NVAA did manage to enlist the support of some of Manchester's businessmen in their campaign to bring down the price of gas in the city.

The campaign for cheap gas
Right up until the first decades of the twentieth century coal-fired steam engines remained the primary source of industrial energy and a major cause of the 'smoke nuisance'. Therefore Lancashire's anti-pollution reformers mainly focused their energies on encouraging manufacturers to adopt the technologies and economical methods that would cut down on waste from using this type of fossil fuel. However, in the 1880s the NVAA began to try to popularise the use of gas, a cleaner and more efficient alternative for providing motive power than 'King Coal'. A gas engine had been invented in 1859 by J.J.E. Lenoir, later developed by Alphonse Beau de Rochas and Nicolaus Otto, to occupy a niche in the market that the steam engine had not filled satisfactorily – providing power to small workshops.[212] The great advantage of employing a gas engine was that it could 'commence as well as cease its action at a moment's notice', doing away with the 'wasteful' necessity of having to maintain a head of steam at all times to operate machinery that was often needed only intermittently.[213] The 'silent Otto' was the most successful gas engine, with Crossley Brothers of Manchester building around 35,000 under license between 1877 and 1900, exporting them all over the world.[214] Two and three horsepower versions of the 'silent Otto' were demonstrated at the Smoke Abatement Exhibitions of 1881 and 1882.[215] Manchester's reformers subsequently urged that the 'substitution of gas engines for steam power' would greatly diminish emissions from the premises of 'some of the worst offenders in the matter of the smoke nuisance', the numerous low chimneys of the city's warehouses where packing presses were utilised.[216] Yet the high price of gas in Manchester, protested the NVAA, was hindering efforts to promote and extend the use of gas-powered technologies.

The municipal authorities at Manchester had long supplied gas to householders, retailers and manufacturers for illuminating purposes, and had long been criticised by consumers for its excessive price.[217] Responsive to the calls of property owners to keep rates low, gas profits were regularly 'expended

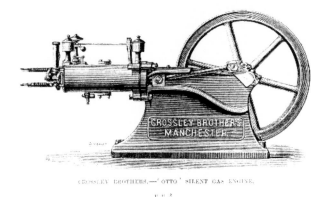

CROSSLEY BROTHERS.—'OTTO' SILENT GAS ENGINE.

Illustration 13. The 'Silent Otto' Gas Engine.

Source: *Report of the Smoke Abatement Committee 1882*, Smith, Elder, & Co., London, 1883.

for the benefit of the ratepayers at large ... in and towards the improvement' of Manchester.[218] The city's anti-smoke activists calculated that no less than £3 million of gas consumers' money had been used in relief of the rates between 1844 and 1911.[219] Businessmen complained angrily that Manchester Corporation's practice of diverting profits from the sale of gas into carrying out public works, such as piping in clean water from the Longdendale Reservoir in the 1850s, kept its cost exorbitantly high.[220] The NVAA concurred; emphasising that this method of keeping down the rates constituted a direct and unfair tax on those who purchased gas:

> ... inasmuch as gas consumers have to pay for improvements ... from which all citizens derive benefit. ... this prohibition of the general use of gas for heating purposes tends to discourage industry [and] enterprise ... and is, in short, a tax upon light, heat, and power.[221]

In the mid-1880s, the NVAA launched a campaign to induce the City Council to lower its price in order to stimulate innovation and the increased use of gas both for motive power and for domestic heating purposes. Gas could make a substantial contribution towards curtailing coal smoke, the reformers insisted, only if the Manchester Gas Committee would reduce its price to the 2 shillings per 1000 cubic feet at which users of power stated it would be 'to their advantage' to use gas engines.[222] If Manchester followed the lead of other towns, such as Hull, Leeds, Newcastle-on-Tyne, and Sheffield, who charged between 1 shilling 10 pence and 2 shillings per 1000 cubic feet, the reformers claimed it would 'enable large numbers of steam users' to substitute gas for coal.[223] But it is important to note that where gas was available at cheaper prices a number of familiar objections to utilising this alternative source of power were soon raised. These included the cost of installing 'complex' new plant, a greater incidence of repairs as compared to coal-fired furnaces, and the 'difficulty of workmen in adapting themselves to changes'.[224]

The narrative of 'waste and inefficiency' performed a prominent discursive role in the reformers' cheap gas campaign. By the early years of the twentieth century, for example, Manchester's reformers were even advancing the idea that it would pay the community to have its gas supplied at cost price. They argued that 'if the ratepayers would forego the profit of £50,000 a year which they make on gas ... they would be going a long way towards doing away with the loss of £700,000 a year which is caused by smoke'.[225] In the end the reformers' campaign did not meet with success. During the period 1880–1911 the price of Manchester gas fluctuated between 2 shillings 3 pence and 2 shillings 9 pence per 1000 cubic feet, never falling to the required 2 shillings mark, while the practice of using the profits in relief of the rates continued unchecked.[226] But, as gas consumption in Manchester outstripped its availability throughout the nineteenth century, the City Council did not face strong market pressure to reduce the price of its gas.[227] Several Manchester businesses supported the NVAA's Cheap Gas Fund; support, however, that came from somewhat predictable sources. The major contributor was Lewis's department store, which had a huge seven-storey building to illuminate.[228] Although more powerful forms of the 'silent Otto' were being developed no widespread demand for gas engines in Manchester is clearly discernible before the First World War, and cheap bituminous coal continued to be the manufacturer's fuel of choice. Indeed, Britain's coal was the raw material for the production of gas. In 1913 just 10 million tons of its coal (or 3.5 per cent of total output) was used to make gas for both domestic and industrial consumption.[229] Additionally, this meant that cheap gas coke was still not being produced in anywhere near sufficient quantities to be extensively used as a substitute for coal in boiler furnaces or in the home. By way of contrast, Britain's industries and homes continued to consume bituminous coal in massive quantities, just over 134 million tons (or 46.8 per cent of total output) and 35 million tons (or 12.2 per cent of total output) respectively in 1913.[230] Alternative sources of power, such as anthracite (the output of which reached 5 million tons for the first time in 1913) and the fledgling electricity industry, increasingly used for lighting purposes, were insufficient and too costly to make significant inroads into resolving the smoke question before the First World War.[231]

As late as 1924, a smoke abatement conference at Manchester heard that a preference for customary coal-fired technologies, especially among Britain's numerous small-time factory owners, continued to be an extremely difficult obstacle to overcome:

> There is no doubt that we are still suffering, especially in the North, from the curse of cheap coal. ... we are conservative folk and the habits of generations are only very slowly outlived. Coal has been, in the past, so cheap that it has not been commercially worth while to bother much about the manner of its burning ... but this is neither all nor the worst of it, the small plant is owned by a manufacturer who specialises not in the manufacture of power but in his own trade. The production of power is to such a one a necessary evil, he does not understand it and often does not wish to; he hires the cheapest man he can get to look after it [the boiler furnace], and as long as his shafting turns round would not give it another thought ... industrialists have been, for the most part, too intent on their own business, and too little mindful of their own and others' comfort, to care how much smoke they poured out of their chimneys.[232]

The NVAA's technological strategies to abate air pollution had proved to be largely ineffective in reducing factory smoke. However, the role of the 'homely hearth' in polluting Manchester's atmosphere had also attracted considerable attention from the reformers.

The home fires
The consumption of coal for domestic purposes in Britain had risen sharply from 12 million tons in 1840 to some 28 million tons by 1887, sparking a lively debate as to whether house or industrial smoke created the most problems in large towns.[233] By the 1880s, some reformers believed that low level emissions from household fireplaces were contributing as much as half of all smoke pollution in Manchester.[234] As well as being irredeemably smoky, as much as 80 per cent of the heat produced by an ordinary open coal fire was thought to 'pass up the chimney with the smoke, and therefore to waste'.[235] The NVAA, unlike MAPS in the 1840s, invested a good deal of time and effort in promoting smokeless alternatives to 'this barbarous system of heating'.[236] Closed stoves, an efficient and economical substitute for the open grate, have already been discussed at some length. All that need be added here is to point out that by the

1900s Manchester's reformers had accepted that deep-seated prejudice against this mode of heating meant that it was 'unlikely that they will be adopted on a sufficiently large scale to have much effect on the smoke problem'.[237] The NVAA, however, vigorously promoted a variety of other appliances for preventing smoke from fires used for heating and cooking.

After mid-century numerous 'smokeless' grates in which coal, coke, or anthracite could be burned came onto the market, often with 'good ventilation' assured. Many such devices were demonstrated with varying degrees of success at the Smoke Abatement Exhibitions of 1881 and 1882.[238] The slow-combustion, underfed grate, which utilised a hopper and a winding device that pushed fresh charges of fuel up into a hot, burning fire, thereby consuming its own smoke, was the most common method. Thomas Horsfall, a leading member of the NVAA, averred that he had used six such 'smokeless' grates economically in his home for many years, although he did admit that they were not very pleasing to the eye.[239] Neil Arnott's grate is perhaps the best known example of this type of appliance, and apart from its 'cheerless' appearance there was another important reason for its lack of success with the general public: its cost. Speaking in 1854, Edwin Chadwick had hoped that the Arnott grate might be supplied to Britain's 'poorest classes' for 'less than one pound each'. He was certain that the grate would then quickly pay for itself as householders not only saved money on fuel, but also lowered their laundry bills as the smoke cloud dissipated.[240] Yet in 1892, probably due to lack of popular demand, the price of an Arnott grate was an expensive £7. 10s. 0d. A sum that Britain's labouring classes could not easily afford.[241]

Smokeless grates were generally to be found in middle-class homes, where heads of households were faced with a similar problem to that of Britain's factory owners: domestic servants often did not use smoke abatement technologies in the way that their designers had intended. The Victorian domestic manual *Cassell's Household Guide* claimed that an underfed grate could save 'an *incredible* quantity of coal ... from 25 to 40 per cent' if only servants would replenish it '*at the bottom*' and refrain from shovelling coal on to the top of the fire in the customary manner.[242] According to Rollo Russell, if correctly handled 'those excellent grates of the under-feeding type would probably abolish two-thirds of the smoke of London ... but no theoretical advantage would prevent a servant from throwing coals on top of the fire instead of pushing them underneath'.[243] Having many other chores to do around the house besides tending to the fire, hard-pressed servants did not ordinarily pay proper attention to good practice in using underfed grates. Where cooking was concerned, the modern kitchen range also promised freedom from smoke and great economy in fuel, as the following advertisement for Marsh's Kitchen Ranges exemplifies.

**NO SMOKE! NO SOOT! NO CHIMNEY SWEEPING!
ENORMOUS SAVING IN FUEL!**

BY THE USE OF THE NEW

SMOKE-CONSUMING AND

FUEL-ECONOMISING

KITCHEN RANGES

AND APPLIANCES

(MARSH'S PATENTS).

Will Burn Night & Day. Will Burn the Commonest Slack

The following Results Guaranteed :—

ECONOMY IN FUEL, 50 to 70 per cent. saved.
ECONOMY IN WORK.
CONSUMPTION OF SMOKE.
EFFICIENCY IN COOKING & BAKING.
EFFICIENCY IN HOT WATER SUPPLY

REFERENCES AND TESTIMONIALS ON APPLICATION.

ADAPTABLE TO

STEAM PANS, CONFECTIONERS' OVENS, and all kinds of FIRES,

Mr. W. T. STEAD in the "Review of Reviews," March 15th, 1892, says :—	The "LANCET," March 5th, 1892, after having appointed a Special Commission to test the claims and merits of this Patent Range, says :—
" I speak of this range as I found it. If the range had turned out badly, I should have said so. As it worked well I have great pleasure in testifying to the fact. If Marsh's Patent acts as satisfactorily when applied to other fires as it has done when applied to my own kitchen range the days of fog are numbered."	" The result is eminently satisfactory, and demonstrates clearly not only the fuel economising effect of the system, but also that the production of smoke is practically nil."

OVER 500 NOW IN USE.

Can be seen Working Daily at

BAXENDALE'S, Miller Street, MANCHESTER.

Illustration 14. 'Enormous Saving in Fuel!'.

Source: T.C. Horsfall, *The Nuisance of Smoke from Domestic Fires*, 1893.

As a rule, advertisers sought to persuade middle-class families to invest in such quality devices, as they were well beyond the means of poorer householders. However, the common work practices of cooks and other kitchen staff often nullified 'guaranteed' benefits. Despite claims to the contrary, most ranges produced a 'great deal of smoke' until they were 'working steadily', particularly as kitchen fires were 'generally forced first thing in the morning to get hot water'.[244] Determined to reduce domestic smoke, the NVAA encouraged Manchester's housewives to install gas appliances rather than 'improved' coal-burning ranges in their kitchens.[245]

Cooking and heating by gas had much to recommend it. Heat was available at the turn of a tap, dispensing with the dirty, arduous, and time-consuming task of laying coal fires. It was economical and efficient inasmuch as gas could be used as and when required, doing away with the need to burn fires all day – especially in the summer months – for cooking and to provide hot water. No harmful smoke and less heat escaped up the chimney and the absence of smoke and soot meant much less labour expended in dusting, washing and cleaning.[246] In addition, by burning coal 'scientifically' in retorts to produce gas, a number of valuable by-products such as coke, tar, sulphate of ammonia and other chemicals used in agriculture, industry and medicine could be recovered, thereby avoiding unnecessary waste.[247] The 1906 edition of *Mrs Beeton's Book of Household Management* confidently asserted that 'there are few persons who have once used a good gas fire that could be persuaded to return to the old method of heating'. Its author was equally as enthusiastic about cooking by gas.[248] From the 1880s on, the NVAA urged householders to 'avail themselves of this convenience, [gas] being economical, cleanly, and effective in the prevention of smoke'.[249] The reformers, however, overlooked the fact that the vast majority of Mancunians could not afford to buy or rent gas appliances, nor could they pay for the expensive fuel to run them. When the NVAA asked the Manchester and Salford Ladies Sanitary Reform Association, working in Ancoats and other heavily polluted districts, if its health visitors would partici-pate in circulating handbills advocating the increased use of gas it refused point-blank to help. The Ladies Sanitary Reform Association was certain that 'it would be useless to distribute them amongst such poor people as their mission women visit'.[250]

In 1884 Manchester's City Council had begun hiring out gas fires and cooking stoves to the public. A short time later, the NVAA pointed out that at Leicester and Sheffield, where gas was supplied more cheaply, sales and rentals of gas appliances were outstripping those at Manchester.[251] The coin-in-the-slot meter introduced in 1890 did help to expand the use of gas in the city. But not until 1903, when Manchester's Gas Committee started to supply and fix gas technology free of charge, did demand begin to grow significantly.[252] However, for most working-class people, who needed to keep a fire burning all day for much of the year to provide warmth, the open hearth was not only 'cheerful' but also the most economical method of obtaining heat. In 1918 Professor William Bone, an expert in the field of fuel technology, estimated that at average pre-war prices of 30 shillings per ton for good house coal, 2 shillings and 6 pence per 1000 cubic feet for gas, and 1 penny per unit for electric current, each unit of domestic heat generated by coal was five times cheaper than gas and twenty-five times cheaper than electricity. Bone concluded that 'for rooms that are in continual use all or most of the day, coal is the cheaper fuel in the long run, whilst for rooms that are used for short periods only, gas-fires are to be preferred'.[253] The barrister C.E. Brackenbury, addressing the Society of British Gas Industries on the topic of air pollution, wasted few words in summing up the situation: 'If we seek deliverance from the smoke nuisance of coal fires ... we must face the fact that it is a question of economics. It is but nibbling at the problem to put a few gas-fires into the houses of the well-to-do.' [254] Yet it would be too simplistic to stress the economic dimension of the smoke question as the sole cause of the slow rate of change from coal to gas in Britain's large towns and cities.

Despite the wholesome claims made for the benefits of gas, its previous widespread use for lighting purposes had shown it to be both dirty and unhealthy.[255] Its sulphurous fumes discoloured and destroyed the fixtures and fittings in people's homes and tainted the air, making it unpleasant to breathe. As well as costing a 'ruinous sum', new gas fires rapidly acquired a bad reputation for making foul smells, giving off insufficient heat, and, crucially, not providing homes with proper ventilation.[256] Furthermore, as gas fires were often badly constructed with ill-fitting flues, dangerous carbon monoxide fumes often leaked into rooms and 'vitiated the air' causing headaches, nausea, and some-times even death.[257] As late as 1909 a meeting of the Society of British Gas Industries was informed that some doctors refused to set foot in a person's home if there was a gas fire burning.[258] The notion that gas heating was not as 'hygienic' as a coal fire took a long time to die away, and only a small proportion of Britain's homes had installed gas fires before 1950.[259] Similar problems attended the gas cooker, with the main difficulty being that 'poisonous' gas fumes were commonly thought to contaminate the food.[260] It took numerous exhibitions, lectures, and a vigorous advertising campaign to finally overcome 'widespread suspicion' about gas stoves, as can be seen from the illustration below. But by the 1930s most households were cooking with gas.[261]

Yet before 1914, despite the much vaunted convenience and cleanliness of gas, many women were still reluctant to abandon familiar methods of heating, cooking, washing, drying and obtaining hot water. To those who were accus-tomed to using one fire for all purposes, the concept of separating the various

Be Modern!

Make your home your castle

The Englishman's home is his Castle—his Castle of freedom. Let it really be free—from dirt, from drudgery and from discomfort.

The Englishman's servant is Gas—which gives his home freedom. Enquire of his wife, who knows all the following truths.

Gas stands for freedom from dirt—no soot, no smoke, no ashes. It stands for freedom from drudgery—no carrying coal, no laying and tending fires, no constant cleaning. It stands for freedom from discomfort—no waiting for a fire to burn up, no badly-cooked unpunctual meals, no wishing vainly for hot water. Where gas fires, gas cookers and gas water-heaters are, you work not, wait not, want not.

Rich man, poor man—both want their homes their castles. With the service of gas, both can achieve their wish; no longer merely castles-in-the-air, their ideal homes become accomplished facts.

GAS FOR
CLEANLINESS, FREEDOM AND COMFORT

Expert information and advice on all gas matters will be gladly given free on application to the Secretary

THE BRITISH COMMERCIAL GAS ASSOCIATION
28, GROSVENOR GARDENS, WESTMINSTER, S.W.1.

Illustration 15. 'Gas for Cleanliness, Freedom and Comfort'.

Source: C. Elliott and M. Fitzgerald, *Home Fires Without Smoke*, 1926.

functions it performed was genuinely alarming. The introduction of modern gas appliances meant radical changes to tried and tested housework routines and practices.[262] Moreover, gas appliances could not be used to incinerate household waste, a drawback that was believed to have serious repercussions for the public's health.[263] It was not cost alone that delayed the adoption of cleaner fuels and 'labour-saving' means of heating and cooking.[264] The grave misgivings that surrounded the new technologies, the open hearth's place as the focal point of home life, and the fact that most people still greatly enjoyed the warmth and comfort associated with a blazing coal fire were equally important explanatory factors. A major impediment concerning the general employment of the gas fire was its 'monotonous uniformity', and, more surprisingly, the fact that it did not 'provide occupation in attending to it and poking'.[265] The scientist Oliver Lodge spoke for the mass of society when he declared:

> We English like an open fire, we like standing in front of it, and toasting ourselves at it; and we can, as a rule, afford to pay for the luxury. Then let us stick to the open fire by all means. It is very cheerful and it ventilates the room well ... all I ask is that it shall not be smoky ...[266]

Victorian science, however, did not furnish a cheap, uncomplicated, smokeless open fireplace suitable for use in the homes of the working classes.[267] But where industrial smoke was concerned, in 1886 no lesser authority than Alfred Fletcher insisted:

> Enough ... has been done to establish the fact ... that smoke is not a necessity, but simply the result either of a bad boiler or a bad stoker, and that all who wish it can now be shown how to make steam without smoke. Thus the question of smoke prevention is no longer a scientific one ... The nuisance is voluntarily imposed ... It must now be dealt with socially.[268]

In the last years of the nineteenth century the NVAA, frustrated by a lack of positive interest in its educational and technological strategies to prevent air pollution, adopted a much more aggressive attitude towards the 'smoke nuisance'.

A change of tactics

Over half a century after the *Manchester Guardian* claimed that MAPS' scientific endeavours had left those 'who persist in poisoning the atmosphere ... without the slightest excuse for their misconduct', the NVAA, as its most ambitious research project neared completion, similarly stated:

> This association has hitherto relied upon moral suasion and educational methods to effect its objects, but experience has shown that there are many offenders who are amenable to compulsion only. Hitherto the latter have been able to plead with some plausibility, that while willing to adopt means for preventing smoke, they did not know what to do, and it would be a hardship to be compelled to

provide costly apparatus, the success of which was doubtful. The work of the Committee for Testing Smoke Preventing Appliances is now practically ended, and the issue of its report in the course of a few weeks will deprive smoke producers of the last shred of cause for continued pollution of the air. The time is therefore appropriate for a determined effort to put down this nuisance which is so fruitful of injury to the public.[269]

For almost two decades the NVAA had tried to achieve its aims by 'conciliatory and educational means rather than by coercive measures'.[270] However, Manchester's manufacturers continually rebuffed the reformers' initiatives and they finally lost patience, opting instead for compulsion in order to protect the city and its people from industrial smoke pollution. In 1894 the Manchester and Salford Noxious Vapours Abatement Association, having shown 'every possible consideration' to 'smokers', was the driving force behind the creation of a new pressure group, the Smoke Abatement League.[271] Unlike MAPS and the NVAA, the League's 'main object ... [was] to enforce the law prohibiting unnecessary pollution of the air by coal smoke'.[272] But before examining the activities of the League in more detail, it is necessary to sketch out a number of influences on the smoke question that had not changed substantially since the 1840s.

Major smoke producers remained influential figures in the industrial towns of late-Victorian Britain.[273] At Manchester, due to its dual role as a commercial and industrial centre, 'hard-headed shopkeepers' and other small proprietors were becoming more prominent on the City Council. But in 1876, when the NVAA was founded, Manchester's big businessmen remained the largest single group on the City Council and they sustained their interest in local politics throughout the rest of the nineteenth century.[274] The region's industrialists also continued to be well represented on the county bench.[275] Lancashire's magistrates were, according to one of their number, 'all either manufacturers or the friends of manufacturers'.[276] Moreover, the fortunes of most of Manchester's and south-east Lancashire's inhabitants, whether they were textile workers, shopkeepers, or mill owners, were still tied directly or indirectly to the success of the cotton industry.

The failure of the NVAA's efforts to induce both businessmen and householders to abate their smoke can also be linked to the unfavourable economic climate in which the society set to work. Lancashire and Britain's previously unrivalled position as the 'workshop of the world' was coming under threat from a combination of strong foreign competition and a protracted economic crisis.[277] Many of Manchester's small cotton businesses went bankrupt during the 'Great Depression' years, as industrial growth and economic prosperity were punctured periodically by a series of slumps extending from the mid-1870s to the mid-1890s. According to one contemporary, in 1878 alone there had been no fewer than 169 failures of cotton firms in the region, with losses amounting to £32.5 million.[278] In the 1890s a number of Manchester's larger concerns agreed upon 'defensive' mergers to ensure their survival.[279] In adverse circumstances, Lancashire's cotton factories reorganised and grew in size to increase their productive capacity. But there was no major investment by cotton manufacturers in innovative new plant. Mill owners, with the cooperation of the workforce, met this challenging situation by significantly improving output through utilising established technologies more effectively, especially in cotton spinning.[280] Increased production, however, was not generally accompanied by improved fuel efficiency. And while social conditions were not quite as bad as in the 'Hungry Forties', for Lancashire's workforce the downturn in trade brought frequent bouts of unemployment, underemployment, hunger and poverty. The circumstances, then, were far from auspicious with regard to persuading the region's industrialists and householders to purchase expensive and uncertain smoke prevention technologies. Additionally, despite some calls for the state to introduce protective duties at this time, deeply embedded free trade, *laissez faire* principles remained the political orthodoxy.[281] As was the case in the 1840s, the authorities did not wish to 'tax' industry with heavy fines for polluting the air, particularly as coal smoke was still closely associated with wealth creation. After mid-century, the Manchester region and other 'industrial reserves' continued to be shielded from costly anti-smoke actions by weak pollution control laws.

Legal responses to smoke

In England the anti-smoke clauses in the Public Health Act of 1875, which consolidated and extended the mandatory provisions of the Sanitary Act of 1866, provided the basis for litigation in most towns until they were amended in 1926. The 1875 Act obliged local councils to prosecute polluters if their furnaces were not constructed so as to consume their own smoke and if their industrial chimneys emitted 'black smoke in such quantity as to be a nuisance'. But it contained the following proviso,

> ... the court shall hold that no nuisance is created within the meaning of this Act, and dismiss the complaint, if it is satisfied that such fireplace or furnace is constructed in such a manner as to consume *as far as practicable*, having regard to the nature of the manufacture or trade, all smoke arising therefrom, and that such fireplace or furnace has been carefully attended to by the person having the charge thereof.[282]

The ambiguous 'best practicable means' rider that weakened the local Acts of the 1840s continued to enable manufacturers to set up an effective defence against prosecutions. This loophole, as John Hassan has observed, albeit in the context of the protection of Britain's rivers, was regarded by contemporaries as 'virtually a license to pollute'.[283] Moreover, no explicit standards relating to smoke emissions were set in the Act, while what constituted 'black' smoke was never defined in the courts.[284] At Sheffield, for example, industrialists loudly chorused that they were allowed to make the 'brown' smoke that issued freely from their chimneys, while London's Chamber of Commerce stated that the word 'black' was 'their only safeguard against more or less vexatious proceedings'. Critics claimed that 'in order to go scot-free ... an offender had merely to make his smoke grey or blue or of any other fancy colour'.[285] The maximum permissible penalty under the Public Health Act of 1875 was an inconsequential £5, but the fines imposed by sympathetic magistrates on offenders were generally much lower. The metals and mining trades were again exempted from its provisions, as were the open coal fires in millions of homes around the country. The legislature continued to compromise with polluting industry so as not to inhibit economic growth. Such discriminating, vaguely worded legislation was widely recognised by contemporaries to be ineffectual as far as curtailing smoke pollution went. The Edwardian barrister C.E. Brackenbury described the 1875 Act as 'hedged round with legal safeguards so as to prevent [factory] smoke being regarded as a nuisance'.[286]

Interpreting and enforcing the regulations
As the statute did not fix any set standard for smoke emissions each local authority was free to decide for itself exactly how much smoke constituted a nuisance. Municipal smoke inspectors used different methods to evaluate the density and opacity of black smoke, and time limits for permissible emissions also varied greatly from place to place.[287] In 1898 a comparison of the time limits for allowable discharges set by Britain's industrial towns revealed the great anomalies that existed in dealing with the 'smoke nuisance'. Manchester City Council allowed only 2 minutes of 'black smoke' in one observation of half an hour's duration, while Salford Borough Council permitted 5 minutes 'dense smoke' in the hour. Other manufacturing towns were far more lenient with Bury, Stalybridge and Wigan allowing ten minutes 'black smoke' per hour, Middleton twelve minutes in the hour, Birmingham fifteen minutes 'dense black smoke' per hour, while Belfast, Burnley, Crewe, Derby and many other large towns set no time limit on emissions whatsoever.[288] Smoke pollution

knows no boundaries, and without the adoption of a uniform low time-density standard Manchester's reformers argued that 'any improvement effected in the city [would] be to a large extent counteracted' by the smoke drifting in from neighbouring boroughs.[289] Setting limits for emissions, however, did not necessarily mean that the regulations would be enforced stringently.

In the last quarter of the nineteenth century Manchester's reformers became more and more disillusioned with the lax way in which local authorities enforced the anti-smoke laws. Salford Borough Council in particular came in for strong criticism, as the prevailing winds carried its smoke into Manchester for much of the year. Despite the surface appearance of zealous activity created by the many thousands of observations taken by Manchester and Salford's smoke inspectors, the number of cases in which firms were actually fined by the magistrates was very small and the penalties imposed were negligible, as Tables 7 and 8 make plain. The fines were trifling in comparison with the penalties exacted by Manchester's Court Leet at the turn of the nineteenth century. As the tables clearly show, the bulk of those who polluted the air simply received a caution from the smoke inspector. But it is important to note that both the Health Committee of the Salford Borough Council and the Nuisance Committee of the Manchester City Council also levied small fines on offenders from time to time. Most notably in the case of Manchester, where the majority of lawbreakers were dealt with by the Nuisance Committee right up until 1887 when the practice ceased. The transfer of 'prosecutions' from Nuisance Committee to the magistrates court accounts for the sharp increase in the number of cases fined after this date in Table 8. Yet, as the annual total of the penalties imposed by the Nuisance Committee exceeded £100 on only one occasion between 1875–88, this does not undermine the substance of the argument presented here. The practice of summoning polluters to appear before Council Committees was employed to save the cost-conscious local authorities the expense of taking offenders to court, and to spare factory owners the indignity of appearing before the magistrates like common criminals. For a time the NVAA attempted to 'name and shame' smoke polluters by providing local newspapers with the identities of those firms that broke the law, but only the *City News* agreed to publish this information.[290] At Salford, where action was often taken under a local Act of 1862, the Health Committee regularly prosecuted the 'negligent stoker' rather than the factory owner. The stoker's nominal fine was nearly always paid by the employer, who nonetheless 'escaped the odium ... of permitting the nuisance'.[291]

Year	Number of observations	Cautions issued by inspector	Cases fined by magistrates	Total amount of fines
1875	n.a.	n.a.	11	£61. 0s. 0d.
1876	n.a.	n.a.	1	5s. 0d.
1877	n.a.	n.a.	1	£2. 0s. 0d.
1878	n.a.	n.a.	1	£1. 0s. 0d.
1879	620	n.a.	27	£43. 0s. 0d.
1880	n.a.	n.a.	n.a.	n.a.
1881	n.a.	n.a.	n.a.	n.a.
1882	453	n.a.	12	£28. 0s. 0d.
1883	598	n.a.	14	£42. 0s. 0d.
1884	573	n.a.	9	£31. 5s. 0d.
1885	540	n.a.	3	£8. 10s. 0d.
1886	524	53	1	£1. 0s. 0d.
1887	411	56	13	£32. 15s. 0d.
1888	1,174	93	11	£20. 3s. 6d.
1889	983	109	22	£49. 15s. 0d.
1890	1,270	64	20	£44. 7s. 6d.
1891	1,485	93	37	£44. 5s. 0d.
1892	1,287	99	15	£23. 0s. 0d.
1893	1,790	112	7	£4. 2s. 6d.
1894	1,715	136	26	£21. 17s. 0d.
1895	2,111	140	15	£19. 10s. 0d.

Table 7. Smoke Observations, Cautions Issued and Cases Fined by the Magistrates at Salford, 1875–95.

Source: County Borough of Salford, *Annual Reports of the Medical Officer of Health, 1875–1895.*

Note: Salford employed just one smoke inspector, who was not appointed on a full-time basis until 1888.

Year	Number of observations	Cautions issued by inspectors	Cases fined by magistrates	Total amount of fines
1875	4,251	342	5	£3. 10s. 0d.
1876	4,080	3,004	15	£8. 0s. 0d.
1877	2,837	1,272	33	£35. 5s. 0d.
1878	3,773	632	n.a.	n.a.
1879	4,224	659	3	£1. 10s. 0d.
1880	4,446	580	7	£3. 10s. 0d.
1881	4,397	576	2	£2. 0s. 0d.
1882	4,395	643	2	£2. 0s. 0d.
1883	4,285	642	5	£5. 0s. 0d.
1884	4,255	623	1	£2. 0s. 0d.
1885	4,238	590	0	0
1886	4,937	759	0	0
1887	6,638	995	26	£22. 10s. 0d.
1888	6,544	985	65	£97. 9s. 6d.
1889	1,436	986	102	£214. 3s. 0d.
1890	788	882	46	£113. 12s. 0d.
1891	944	1,197	80	£139. 6s. 0d.
1892	1,472	1,611	38	£67. 11s. 6d.
1893	2,000	2,244	63	£97. 12s. 0d.
1894	1,782	1,933	66	£119. 19s. 6d.
1895	1,797	2,002	98	£180. 19s. 0d.

Table 8. Smoke Observations, Cautions Issued and Cases Fined by the Magistrates at Manchester, 1875–95.

Source: City of Manchester, *Proceedings of the Council, 1875–1895.*

Notes: Manchester employed two full-time smoke inspectors up to 1885, when a third was taken on. A fourth inspector was appointed in 1890. The great number of cautions issued in 1876–77 is a response to the introduction of the anti-smoke clauses in the Public Health Act of 1875. The fall in numbers of smoke observations after 1888 does not represent a lessening of activity on the part of smoke inspectors, rather a more rigorous approach to gathering evidence as all offenders were dealt with by the magistrates from this date.

Further examination of the tables also shows that the yearly movement in numbers of prosecutions broadly followed the trade cycle, a trend that Robert Gray has recently identified regarding Factory Act proceedings.[292] The worst years of the 'Great Depression' period were 1877–79, 1884–86, and 1891–93 at which time the numerous infringements of the smoke regulations in Manchester and Salford were treated even more leniently than was usual. Additionally, in 1882 Manchester City Council had included a clause in a local Act that allowed its magistrates to fine offenders a maximum of £10 per day for offences against the smoke regulations.[293] Table 8 confirms the NVAA's complaints that these 'increased powers ... [were] entirely in abeyance'.[294] Over the two decades 1875–95 the local authorities merely 'winked at the pollution of the atmosphere', rather than take meaningful action to discourage the production of industrial coal smoke;[295] although Manchester City Council perhaps deserves some credit for hauling indignant local manufacturers before the magistrates in greater numbers after 1887.

In spite of the setbacks from cyclical depressions and worries about foreign competition, between the mid-nineteenth century and the First World War the cotton industry continued its impressive growth. Although Britain's market share of world cotton goods peaked at 82 per cent in 1882–84, in absolute terms output was at its highest in 1913 when an 'all-time, all country record of almost 7.1 billion yards of cotton cloth' was sent for export.[296] The lion's share was manufactured in south-east Lancashire. Increased output and relatively high wages meant that in the generally prosperous conditions of the late nineteenth century there was an 'upward trend in most aspects of living standards' for the region's workforce.[297] As the hub of the world's cotton markets, the weekly turnover of trade in Manchester had increased 'from £1 million in the 1850s to £10 million by the 1880s'.[298] No wonder, then, that some contemporaries were outraged at the indulgent administration of the anti-pollution laws and the insubstantial fines exacted by Lancashire's magistrates. In the letters columns of the *Manchester Guardian* 'Sanitas' questioned the wisdom of continuing to pay smoke inspectors to collect 'information which is not utilised', while 'Observer' complained, 'What is a fine of twenty or forty shillings to a concern turning over millions per annum?'[299] At the turn of the twentieth century the Reverend W.E. Edwards Rees vehemently protested,

> ... there are people in Manchester who systematically pollute the air, and systematically pay the fine – the ridiculous fine that is imposed: it is a mere bagatelle. It is much cheaper to pay the fine than to put up new plant.[300]

Legislation that was 'insincerely drafted and meant' had little or no effect in cutting atmospheric pollution.[301] Nor did it encourage the development and use of more efficient smokeless technologies. In 1894, dissatisfied that many industrialists were stubbornly refusing to take effective measures to prevent smoke pollution, Manchester's reformers resolved to take the law into their own hands.

The Smoke Abatement League

The legal actions of the authorities at Manchester and Salford had achieved little by way of a visible improvement in air quality. But in the great majority of Lancashire's factory towns such as Ashton-under-Lyne, Oldham, Rochdale, Stalybridge, and Stockport the smoke clauses in the Public Health Act of 1875 had seen even less service.[302] In many manufacturing towns they were simply 'not operative', as town councillors insisted that 'the discouragement of smoky chimneys may mean a blow to local industry'.[303] Heavy smoke emissions from the unencumbered tall chimneys of neighbouring towns exacerbated Manchester's air pollution problems, and not only in a physical sense. Manchester's industrialists drew attention to 'the greater leniency and discrimination shown by municipal authorities elsewhere towards their staple industries'.[304] They complained that they were being unfairly treated when works situated beyond the city's boundaries generally escaped regulation. In 1894 a deputation from the Chamber of Commerce threatened the City Council with the wholesale 'remov[al of] their works away from Manchester, with a consequent loss to the city of thousands and thousands of pounds in rates, and loss of employment for thousands of our workpeople', unless it stopped 'harassing' industry for sending out coal smoke.[305] The Local Government Board, a central body that could have intervened over the non-enforcement of anti-pollution laws, was unwilling to force local authorities elsewhere in cotton Lancashire to take action against the 'smoke nuisance'.[306] Since local self-government was the cornerstone of the Victorian political system, Parliament and the Local Government Board 'had always to work within the limits of consent' on major issues.[307]

    Stung by the Chamber's blatant attempt 'to claim indulgence for their trades', and aware of the pressing need to change smoke control policy in Manchester's satellite towns, the NVAA accelerated its plans to form a more extensive Smoke Abatement League. At a conference convened at Manchester by the NVAA in December 1894 it was decided that 'steps be taken to form a Lancashire Smoke Abatement League, with branches in the principal towns in the County'.[308] The following year the Manchester and Salford Noxious Vapours Abatement Association transformed itself into the Manchester and Salford Branch of the Smoke Abatement League, which was to have its headquarters in Manchester. Its primary goal was to enforce the smoke abatement laws rigorously and comprehensively.

Any discontented citizen, under the provisions of section 105 of the Public Health Act of 1875, could take proceedings against manufacturers who allowed their chimneys to emit black smoke. The League intended to pursue private prosecutions through the courts in cases where the authorities had failed to take action against persistent lawbreakers. Indeed, Manchester's reformers harboured grand ambitions to extend the League's activities onto the national stage.[309] However, the Manchester and Salford Branch of the new Smoke Abatement League was not strongly supported by the public, and it did not receive sufficient donations to carry out its intentions as planned. In any event, it did not prosecute a single case. From the outset the Manchester and Salford Branch was struggling to keep afloat on a meagre income of £30 to £90 per annum, not nearly enough to conduct a costly legal campaign.[310] The Manchester-based League's affiliated branches at Sheffield and Middleton did pursue a more active policy with regard to prosecutions, but without any marked success. The financial records of the League's Sheffield and Rotherham Branch show that they expended the sum of £60 13s 11d on 'Counsel's Fees, Witnesses, etc.' in 1895 in order to take action against five manufacturers; just three were convicted incurring fines of only £3 3s 6d between them, including costs. The costs awarded to the Sheffield Branch by the magistrates amounted to £2 10s 6d.[311] To make matters worse, the League's work in Sheffield was suspended in 1897 when the evidence of its smoke inspectors, who were untutored in 'the thousand and one difficulties of practical smoke prevention', was not accepted by the courts.[312]

The Middleton Branch of the Smoke Abatement League was the undertaking of a single individual, Mrs Susan Fanny Hopwood.[313] Between the years 1892 and 1900 Mrs Hopwood, at her own expense, took around 400 offenders to court in Bury, Middleton, Oldham, Rochdale and other Lancashire cotton towns.[314] While she was successful in most cases, the magistrates in these districts usually imposed fines of less than £1 on the guilty parties. The returns on her considerable outlay were negligible, as Mrs Hopwood tacitly acknowledged in her annual report for 1900:

> I am sorry to have a very unfavourable report concerning a few of the Middleton Manufacturers. Two of the richest firms in the borough have had to answer for close upon twenty summonses each during the year, for an average duration of seventeen minutes of black smoke per hour; and considering that it is over eight years since the firms were ordered by the bench to abate their smoke, in accordance with the Health Act, we are travelling but very slow at Middleton.[315]

At the same time, the Sheffield and Rotherham Branch of the League was protesting in its annual report that the 'Public Health Act, as interpreted by the local magistrates ... is a farce'.[316] Like the Reverend Molesworth before them, the League's activists recognised that if the smoke question was to be solved it must be removed from the local domain, where 'the great hindrance to effective administration of the law ... from long experience [was] found to be the result of the strong representation on Local Authorities and the Bench, of those who create the nuisance'.[317]

In 1881 Professor A. Hopkinson, an authority on Jurisprudence and Law at Manchester's Owens College, had recommended in a report prepared for the NVAA that smoke pollution be placed under the jurisdiction of the Alkali Inspectorate.[318] Two years later, the society's Thomas Horsfall told a meeting of the National Smoke Abatement Institution in London that he hoped for

> ... the appointment of one or two Government smoke inspectors for our large towns. This matter ought not to be left in the hands of persons who, if they are not themselves interested in the production of smoke, are so surrounded with friends who are, that it is quite impossible that they should take an unbiased view of the subject.[319]

To have the 'smoke nuisance' assigned to the control of a strong centralised inspectorate, which would ensure that uniform action was taken across the whole of the country, was a long-standing and important goal for many of Manchester's reformers. After 1895, the Smoke Abatement League continued to lobby for the regulation of smoke emissions to be placed in the hands of technically competent government officials. During the years 1896–97, the League worked alongside the Smoke Abatement Section of the Leeds Sanitary Aid Society preparing an 'influentially signed' petition to the Local Government Board to this effect. Both pressure groups urged that either the Alkali Acts be extended to cover all factory chimneys or a 'distinct staff of Government Inspectors, who would be independent of local influences' be created.[320] The President of the Local Government Board, however, did not consent to formally receive their petition and no steps were taken to wrest control of the smoke problem away from local authorities.[321] The well-organised opposition of the industrial lobby in Parliament to tighter regulation undoubtedly had a hand in this 'diplomatic brush-off'.[322] However, another factor that militated against the authorities taking a tougher stance toward industrial smoke was that British householders could still pollute the air with impunity.

The Englishman's castle

No government of the period was willing to incur the public's displeasure by taking action to interfere with their freedom to enjoy the warmth of a traditional open coal fire. Parliament, broadly in favour of *laissez faire* principles, was reluctant to legislate against private households for 'If an Englishman's house is his castle, the domestic fireplace is the keep of the castle, the very centre and citadel of the stronghold.' [323] Yet emissions from domestic chimneys were recognised to be a major source of air pollution, even in industrial districts such as Ancoats where workers' homes were overshadowed by smoky mills and factories.[324] The reformers were afraid that city air would continue to be heavily polluted with industrial smoke so long as domestic emissions went unpunished. Firstly, because a high proportion of the urban smoke cloud was attributable to the traditional open fire, it was possible to argue that it was 'not worth while to impose on manufacturers the trouble and cost of preventing [their] part of the nuisance'.[325] Secondly, as householders were exempted under the Public Health Act of 1875 it afforded polluting industrialists the opportunity to claim that they were being unfairly victimised. As Thomas Horsfall pointed out, 'it is very discouraging for manufacturers to find such a large amount of smoke coming from the houses ... the manufacturer not unnaturally thinks that he is very unreasonably dealt with if he has to go to a very large expense to diminish his share ... of the smoke'.[326] That the smoke that issued from household fires was widely perceived to be 'non-productive' did not improve matters. Domestic smoke emissions were commonly associated with comfort and not with wealth-creation: a representation that hugely devalued the arduous labour of millions of women. Nevertheless, while the 'non-productive' smoke from a domestic chimney was immune from prosecution, local authorities and the state were unwilling to put the full force of the law in motion against the 'productive' smoke of the factory owner. According to Thomas Horsfall, only a new law – 'strictly enforced' – prohibiting domestic smoke would finally clear the air of the factory towns. For, as Horsfall himself admitted, 'I believe that unless this cloud can be made, like Wordsworth's cloud, to move all together, it will not move at all.' [327] However, Britain's anti-smoke reformers did not envisage that legislation 'of a very drastic and revolutionary character' would soon be forthcoming.[328] As domestic smoke prevention technologies were expensive, they mostly accepted that little could be done 'to force the poor cottager to cease burning his scanty handful of coal in the manner and under the conditions to which he has been accustomed'.[329] In the meantime, all the League's activists could do was to call upon public-spirited citizens to 'voluntarily [make] their homes or the houses they own smokeless ... For each such house proves to many persons, who but for it would not know, that house-smoke can be prevented'.[330] There was, however, no noticeable response from the public to the reformers' appeals for them to 'do their bit' for the community and the environment.

Prominent 'anti-smokers' themselves tried to set an example for the public to emulate. For example, Thomas Horsfall, Dr H.A. Des Voeux,

Treasurer of London's Coal Smoke Abatement Society, and George Davis, District Inspector of Alkali Works, all lectured and wrote on the smokeless methods they employed for heating and cooking in their own homes.[331] They urged Britain's local authorities to likewise set a good example by installing underfed grates and other smokeless appliances in Town Halls, libraries, schools and new houses. In addition, reformers also tried to encourage householders to change their coal-consuming lifestyles by dispelling the notion that small, individual contributions to the 'smoke nuisance' were of no real consequence. Glasgow's Chief Sanitary Officer, Peter Fyfe, made a point of broaching the subject in a public lecture on the effects of smoke pollution. He told the assembled crowd, 'I doubt if any of us, householder or manufacturer, ever dreams, when he hears of our appalling infant mortality, that he has a vital share in bringing it about when he or his contributes to the befouling of the air.' [332] Similarly, at Manchester, George Davis enquired of those attending an anti-smoke meeting:

> I ask you present here to-night, has any one of you done anything to produce a smokeless chimney in your own house, or to help in any way to render the air of this great city any purer than it was a year ago? ... I think I am correct in saying that not one householder in ten thousand, by the emission of it [smoke], ever thinks for a moment that he is an operator in the great Air Pollution Company. The sooner the householder understands the real bearings of this question, the nearer we shall be to the time when so-called 'fogs' will be numbered among the things of the past.[333]

As it was 'impossible to draw an indictment against a whole people', education, not coercion, was thought to be the key to abating domestic smoke.[334] But the reformers' technical advice and their numerous appeals to self-interest and economic rationality made little impression on the Victorian public. The exertions of the NVAA and the Smoke Abatement League had largely failed to unravel the densely woven cultural fabric that had created a positive image of Manchester's coal smoke. And it is worth noting that in 1909 the chimneys of Manchester's Town Hall were the subjects of a smoke nuisance complaint.[335]

The NVAA and the Smoke Abatement League in brief

First as the Noxious Vapours Abatement Association and latterly as the Smoke Abatement League, Manchester's reformers campaigned vigorously for the control of air pollution for almost three decades. Science and technology did offer solutions to the smoke question, as the NVAA's public demonstrations and detailed investigations helped to establish. But the confusing number of prizewinners at the London and Manchester Exhibitions, which had in any case achieved success under questionable conditions, did little to dispel uncertainties about 'smokeless' appliances. Furthermore, the dense statistical data concerning the performance of innovative smoke prevention technologies working *in situ*, amassed by the experts of the Committee for Testing Smoke Preventing Appliances, were not easily understood by manufacturers who were neither engineers nor proficient in the science of combustion. Inflated claims about the 'enormous savings' to be made on fuel bills, which often failed to materialise both in the home and in the workplace, only served to confirm suspicions that smoke prevention could not be made into 'a paying business'. However, it was possible and profitable to operate even the most basic of steam technology without much smoke if sufficient boiler room was provided and the stoker carefully attended to the furnace. But, like the earlier Manchester Association for the Prevention of Smoke, the city's late-Victorian pressure groups did not always clearly signpost the pathways to cleaner air.

To Manchester's industrialists, primarily concerned as they were to increase production, the efficient combustion of cheap coal was unarguably a low priority. Few businessmen actively supported the reformers' initiatives. What is more, they strongly resented even the small fines imposed by the local authorities when householders and works situated beyond the city's boundaries went unpunished. In 1889 Harry Grimshaw, magistrate, manufacturer, and member of the City Council, who asserted that only 20 per cent of Manchester's smoke-cloud was caused by industry, informed a public meeting on air pollution:

> A certain quantity of coal had to be burnt, and as a natural result smoke was produced ... It was an easy thing to drive away industries from one centre to another, and it behoved Manchester people, if they were not indifferent to their own interests, to be just and candid in the matter of the smoke nuisance. They must endeavour to clear the city atmosphere without harassing ... trade almost to the verge of extinction. ... Corporations should have consideration for the industries that had made our cities and towns ... [336]

Most of Lancashire's manufacturers believed that they had fulfilled their obligation to reduce urban air pollution once they had erected a tall chimney to transport their smoke away from the town. Furthermore, as Grimshaw's remarks make plain, the distress call 'industry in danger' was still doing good service at the end of the nineteenth century.

Despite creating the illusion of meaningful activity, the authorities at Manchester and Salford made no 'real efforts' to remedy the 'smoke nuisance'.[337] In the boom year of 1882, Alderman Isaac Bowes, industrialist and influential member of the Salford Health Committee, addressing a public meeting on air pollution, provided a useful insight into its policy towards industrial smoke:

> ... the Committee had to consider the interests of manufacturers and of the working people who were dependent upon them. The question of the emission of dense smoke was one of degree, and if they were to prohibit it altogether where were they to draw the line? Other smoke, quite as much as that emitted from the chimneys of large works, polluted the atmosphere ... He did not at all coincide in the view ... that heavier fines should be imposed upon offending manufacturers, as he thought they had enough to contend with already ... The Corporation had a double duty to perform, for they had to look after the health of the inhabitants and also to see that struggling industries were not overweighted.[338]

The language of social-cost balancing is to the fore as Bowes implicitly argued that the benefits to health of strictly applying the law were outweighed by the jobs and prosperity that smoky industry provided. And the labouring classes of the twin cities were, for the most part, willing to accept this pragmatic trade-off. Low fines had effectively given a 'quasi-immunity' to Manchester's polluting industries. For no town, as Anthony Wohl has observed, was about to rigorously enforce legislation that 'would drive industrialists to relocate in districts where more lax attitudes prevailed'.[339]

Having usurped the role of the local authorities as the 'protectors' of the urban poor and the urban environment, the Manchester and Salford Branch of the Smoke Abatement League and its associated offshoots enjoyed little success in compelling northern industrialists to reduce their smoke. The bench was, as ever, sympathetic to manufacturing interests, while the League lacked the substantial financial resources necessary to act as an effective enforcement agency. In a *laissez faire* climate, the Local Government Board, which could have taken legal proceedings where councils were inactive, was more interested in tutoring local authorities in the principles of sound administration than in imposing solutions to pollution problems from above.[340] During the Edwardian period the reformers' campaign to clear Manchester's air stalled. Unsuccessful in its attempts at coercion and with a dwindling subscription list, in 1904 the Manchester and Salford Branch of the Smoke Abatement League merged with the Manchester and Salford Sanitary Association.[341] The Sanitary Association continued to press for the 'smoke nuisance' to be placed under central control. But, as in the mid-Victorian period, air pollution was seldom its foremost concern. Without the wholehearted support of the community that they were trying to protect, the NVAA and the League – other than raising public awareness that 'dirt was not cheap' – largely failed to achieve their goals.

Manchester's smoke problem was not resolved before the First World War and its air was ranked among the most heavily polluted in the country until the development of smokeless zones in the 1950s. But how that process began is another story.

Epilogue

Too Little, Too Late?

Towards the end of the First World War, Thomas Horsfall, the veteran anti-pollution campaigner, launched a vitriolic attack on the people of Manchester, who had conspicuously failed to lend their support to the smoke abatement movement. Speaking on the topic at the annual meeting of the Manchester and Salford Sanitary Association in July 1918, Horsfall did not pull any punches in complaining that he and other activists had been 'working for an ignorant public that had been dismissed from school at too early an age to take an intelligent interest in the laws of physical and moral health'.[1] Another commentator was equally scathing in arguing that public inaction was playing a major role in perpetuating the smoke problem:

> The general public have suffered air poisoning for centuries without a protest, and they richly deserve all the disabilities and damage received therefrom. Had they protested and demanded that everything should be done that was practicable to prevent air poisoning, and threatened Members of Parliament that unless they did it, votes would be given to others who would, action would have been taken years ago. Parliament, receiving the public mandate, would have passed the necessary legislation and the Ministry of Health would have demanded that the local authorities should put in motion all the administrative machinery. But this has not been done, and the greatest of all the defaulters are the general public who have, on this great health question, been appallingly apathetic.[2]

To frustrated reformers such as Horsfall, who by 1918 had been campaigning against the 'smoke nuisance' with little success for over forty years, the desire to brand most townspeople as ignorant or apathetic, or both, in order to explain away the lack of widespread support for the movement is understandable. After all, Manchester's reformers – and Horsfall in particular – had expended a considerable amount of time, effort and money over the years in propagating the message that smoke pollution was harmful 'matter out of place'.

From the early 1840s onwards Manchester's anti-pollution activists had stressed that coal smoke was the cause of the needless destruction of both green spaces and monumental architecture. They argued that it was a costly tax that ate into the weekly budget of every urban family by injuring clothes and possessions and adding to the arduous tasks of washing and cleaning. Activists also attempted to substantiate direct links between air pollution and the large-scale loss of life from chronic respiratory diseases. In addition, they insisted that smoke was an unnecessary and 'unscientific' waste of energy and fossil fuel. Finally, in the last quarter of the nineteenth century, smoke was depicted by campaigners as a potent source of physical and moral degeneration among Lancashire's urban poor. Yet the reformers' frequent appeals to economic self-interest, civic and national pride, moral and humanitarian principles, and for the conservation of finite natural resources attracted only very limited, mainly middle-class support for Manchester's smoke abatement movement. Not, however, because working-class Mancunians were indifferent where coal smoke was concerned, or because they were simply too dull-witted to grasp the often complex arguments of the 'anti-smokers', but because the vast majority of the labouring poor afforded a higher priority to ideas that associated this form of air pollution with wealth-creation and personal well-being.

For most people living in the great northern 'towns of labour' coal smoke was not readily perceived as dangerous 'matter out of place': they saw it instead as a 'natural' part of the urban-industrial landscape. While to many reformers smoke emissions were the antithesis of progress and civilisation, to Lancashire's workforce this type of air pollution was commonly understood to be good, honest dirt. Manchester's omnipresent smoke cloud could not easily be ignored, and by the 1880s over a century's shared experience of living with its belching chimneys had created a powerful collective belief that air pollution denoted employment and 'good times'. Over the course of several generations a unity of outlook was fashioned by its inhabitants' active involvement in polluting the atmosphere during mundane everyday activities, and through the construction of images and narratives that portrayed the city's 'inevitable' coal smoke in a positive light. Smoking factory chimneys were generally interpreted as a reassuring sign that the enterprising Manchester region, the engine room of the 'workshop of the world', was still prospering. This interpretation was under-lined by the communal hardships experienced by urban dwellers at times when the air was clear, during trade depressions and strikes. Throughout the Victorian and Edwardian periods sustained economic growth was the overriding priority for factory operatives and factory owners alike. Cotton Lancashire's inhabitants lacked enthusiasm for the reformers' initiatives to curb air pollution because they thought that stringent regulations and uncertain smoke abatement devices might damage productivity and endanger their precarious livelihoods.

This broad consensus in favour of the pursuit of economic prosperity rather than the protection of the physical environment was reflected in changes to the ways that the laws relating to air pollution were interpreted and enforced. As Manchester went 'Steam Mill Mad' pre-existing regulations that had allowed its authorities to impose heavy fines on smoke polluters were relaxed to facilitate the rapid growth of industry. Almost all subsequent legislation introduced to combat smoke – both local and national measures – included an ambiguous 'best practicable means' clause and imposed insignificant fines that, taken together, effectively provided Victorian Britain's industrialists with a license to pollute. The case is often made that it was all but impossible for the authorities to take tougher action against polluting businesses, owing to the predominant doctrine of *laissez faire* and the lack of widespread popular support in favour of smoke abatement. By the same token, the traditional argument emphasises that a 'slow, piecemeal, persuasive' approach to pollution control, rather than compulsion, was the only viable strategy available to public officials working to bring cleaner air to Britain's cities. Eric Ashby and Mary Anderson, for example, make something of a virtue out of the pragmatism, ad hocery, and endless patience demonstrated 'in nearly two centuries of evolution' of British environmental policies.[3] Anthony Wohl, for his part, asserts that it 'would be whiggish' to view Britain's inadequate legislative response to air pollution too critically as 'a case of too little too late'. He reinforces the notion that given the difficult social, economic and political circumstances, the only practical option open to the authorities at this time was to call for the voluntary cooperation of industrialists so that smoke abatement could 'proceed at a reasonable pace'. Indeed, for Wohl the patient endeavours of the Victorian state in attempting to 'clean up the skies' are important first steps towards protecting Britain's environment.[4] Tapping into a similar vein, Catherine Bowler and Peter Brimblecombe's recent study of air pollution in Manchester tells a tale of a forward-thinking municipality continually striving to develop a coherent policy to reduce its smoke emissions.[5] Yet before the First World War smoke abatement had made little progress in this 'pioneering metropolis', nor in any other northern industrial town. At best, to use Wohl's oft-quoted words, all that was accomplished was 'to turn the urban skies of Britain from a gritty black to a dull grey'.[6]

Nevertheless, a common thread running through these studies (and many others) is the view that the gradualist approach to pollution control has been something of a success story overall. For this school of thinkers, the slow, incremental, and haphazard progress of government policy to regulate smoke emissions is seen as an effective way of combating air pollution, that from inauspicious beginnings steadily advanced to culminate in the more stringent environmental protection laws passed in the later twentieth century. In contrast, in the present book, I stress that local and central government bodies could have done far more to improve air quality in Victorian Manchester. However, at a time when the science of smoke prevention was becoming better understood, the laws prohibiting air pollution were weakened rather than strengthened, slowing the adoption and development of innovative, cleaner technologies. Furthermore, the increasing use of the doctrines of 'prior appropriation of land' and 'social-cost balancing' by Britain's judiciary resulted in heavy industry becoming concentrated in specific districts, such as factory-dominated Ancoats, where polluters could be shielded from 'ruinous' law suits.[7] In the nineteenth century, the Manchester region emerged, to paraphrase Franz-Josef Brüggemeier, as an area where industry rather than the urban environment or the people's health was consciously protected.[8]

The air of Victorian Manchester was so severely polluted that many thousands of its inhabitants suffered from chronic lung conditions, rickets, and other 'diseases of darkness'. Indeed, by the last decades of the century the bronchitis group of respiratory diseases had unobtrusively become the 'most important single killer' in England's factory towns.[9] The industrial zone of Ancoats, which was home to over 55,000 of Manchester's working people by 1861, recorded some of the country's highest levels of air pollution as well as some of its highest death rates from respiratory disorders.[10] However, while the mill hands, artisans, and unskilled labourers of 'Cottonopolis' had only limited opportunities to escape the health-impairing coal smoke, it does not necessarily follow that they were the helpless victims of environmental pollution. We must remember that by the 1840s south-east Lancashire's working classes were a significant political force, not lacking the muscle or the will to challenge both large employers and Westminster to protect their standards of living and oppose the corruption of central government. Indeed, the history of the Manchester region's working classes is one of men and women who were 'fired with enthusiasm for changing their world'.[11] To enhance their quality of life they fought for the People's Charter that aimed to give the majority a voice in government; they were actively involved in Owenite socialism and the co-operative movement; and they organised in trade unions to sustain wage-levels and improve conditions in the workplace. Moreover, as the culture of radicalism gained strength in cotton Lancashire, Manchester inevitably became the centre for mass meetings whose scale genuinely alarmed those in authority. For example, it was estimated that a crowd of over 250,000 people attended a Chartist meeting held at Kersal Moor, Manchester, on 25 September 1838, mobilised from numerous towns throughout the textile districts.[12] Thus, if the industrial communities of Lancashire had been determined to eradicate air pollution, there can be little doubt that they could have lent strong support to the embattled smoke abatement movement.

However, while working-class protests to improve wages and working conditions were commonplace during the nineteenth century, where the production of coal smoke was concerned employer and employee generally saw eye to eye. The interests of Lancashire's manufacturers and workforce meshed, both groups subscribing wholeheartedly to the notion that 'wheer there's smook there's brass'.[13] Against the backdrop of cyclical economic depressions and increasing international competition for markets, most working people were inclined to side with local mill owners when they insisted that smoke control was 'not a paying business'. As a result, in June 1891 Reginald le Neve Foster, prominent industrialist and director of Manchester's Chamber of Commerce, could argue convincingly that the efforts of the region's anti-smoke activists were 'neither ... appreciated nor supported by the public'.[14] And in the years from 1837 to 1910 it is true to say that the industrial communities of the Manchester region did not show much enthusiasm for smoke abatement. Working men and women listened to the arguments put forward by reformers, but despite being most acutely affected by polluted air they never turned out in force to demand a cleaner atmosphere. Yet this does not indicate that the labouring poor were indifferent to the reformers' attempts to curb smoke emissions. On the contrary, they were willing to assert their views on the subject very vigorously indeed. For example, factory workers regularly hurled abusive language at the Smoke Abatement League's Mrs Susan Fanny Hopwood as she made her way to court to prosecute offending manufacturers. They further demonstrated a lack of appreciation for Mrs Hopwood's actions by pelting her carriage with rotting fruit, vegetables and 'market filth'.[15] Working-class hostility towards the anti-smoke campaign was also apparent at Sheffield, as the veteran reformer Edward Carpenter recalled in 1921:

> More than thirty years ago our little group of Socialists in Sheffield – and we are proud of the fact – set about urging this very question [smoke abatement] on the consideration of the City Council and the local population ... We got a table out into the street, made speeches, and distributed leaflets and pamphlets. The work-people mostly jeered. 'They want us to do without smoak,' they said, 'but how can we live without smoak?' 'If there were no smoak there'd be no tra-ade!' This sounded convincing; and the argument, such as it is, is even now repeated when present-day reforms are proposed.[16]

Although many working people were determined to fight to build a better life for themselves in Britain's manufacturing towns, they long remained unconvinced of the merits of smoke control. To most Mancunians coal smoke meant jobs and money, and it is hardly an exaggeration to say that they viewed its absence in the urban landscape as nothing short of disastrous. Moreover, this was not the only obstacle that Victorian and Edwardian anti-smoke societies faced in terms of harnessing the weight of popular opinion to their cause.

There was a widespread reluctance among townspeople to give up the smoky open hearth, a technology that had become central to everyday home-life in Britain. The traditional coal fire was not only an economical choice of home heating, it also denoted warmth in every sense of the word, not forgetting its important practical uses in cooking hot meals, providing boiling water, ventilating crowded rooms, and incinerating household waste. No government of the day was willing to risk interfering with the rights of millions of Britons to enjoy a 'cheerful' open fire in their own home for fear of incurring the public's wrath. Not until London's protracted 'Great Smog' of December 1952 brought about the premature deaths of some 4,000 people did the government feel that it had sufficient public support to bring in tough new legislation that, for the first time, covered domestic as well as industrial smoke emissions. Public opposition to smoke control largely evaporated as the sharp increase in mortality from bronchitis, pneumonia, and heart failure (caused by respiratory distress) finally led to the realisation that this form of air pollution could be just as deadly as the cholera epidemics of the nineteenth century.[17] During the two decades after the passage of the 1956 Clean Air Act the skies finally began to clear over Manchester, London, and other large towns and cities. The irony is, as John McNeill has recently noted, that from the mid-1950s cleaner air allowed more sunlight to penetrate to the streets of formerly smoky cities, where it reacted with pollutants emitted from vehicle exhaust pipes to form dangerous photochemical smogs.[18] Recent research by government scientists concerning motor vehicle emissions suggests that this type of pollution 'brings forward' the deaths of 10,000 people each year in Britain by exacerbating existing heart and lung problems.[19] Tall chimneys belching coal smoke no longer dominate in the urban landscape, but fresh air is something that still remains out of reach for many of the nation's most vulnerable city dwellers; not least of all in early twenty-first century Manchester.

Making connections

Today, rather than local smoke and soot emissions, it is the less visible pollutants from burning fossil fuels that are a major source of pressing global ecological problems. The atmospheric build-up of carbon dioxide and other 'greenhouse gases', produced by power plants, factories, and motor transport, pose a serious threat to the world's environment. Evidence from the natural sciences strongly suggests that 'global warming' will result in a whole range of adverse impacts that include desertification, droughts, species extinction, sea-level rise, and a destructive change in weather patterns that threatens the prosperity, health, and security of both present and future generations. As concern mounts over global climate change, the new discipline of urban environmental history can provide an invaluable perspective on how and why harmful air pollutants can become a 'natural' part of a shared industrial culture. By looking back at how pollution problems unfolded in the past, we may better understand the challenges we face in solving the socio-ecological dilemmas of the present. To conclude, I would like to briefly signpost a number of areas where this case study of nineteenth century Manchester, the spiritual home of air pollution, might usefully help to put 'current problems in fuller perspective'.[20]

First of all, while Victorian and Edwardian ideas and attitudes concerning 'good airs and bad' are, of course, in many respects very different to our own, some ways of thinking about atmospheric pollution have not changed substantially since the onset of the Industrial Revolution. There are, for example, striking parallels to be drawn between the alarmist claims of the pro-smoke lobby in nineteenth century Britain and the doom-laden rhetoric of the fossil fuel lobby in contemporary America, the world's greatest consumer of energy. The Global Climate Coalition, an organisation that represents 6 million U.S. businesses, including the mining, oil, and coal industries, issued the following press release in the wake of the Kyoto agreement of 1997 to reduce greenhouse gas emissions:

> The global climate agreement approved by U.S. negotiators today will send 2 million Americans to the unemployment line in the first ten years, drive up energy costs and consumer prices, and drain at least $150 billion a year from the U.S. economy … 'Our future is being handed over to our international competitors.'[21]

Eugene Trisko, speaking for the United Mine Workers of America, strongly supported the Global Climate Coalition's position on the Kyoto agreement. According to Trisko, if the Kyoto Protocol became a legally binding treaty it would mean 'The 13 million members of unions all across America have reason to fear for their economic futures and that of their children'.[22] Effective lobbying by the U.S. fossil fuel industry and labour unions is at present delaying international negotiations to combat climate change, and whether crucial cuts in global carbon dioxide emissions will ultimately be achieved remains very much in the balance. What is clear, however, is that narratives associating industrial air pollution with jobs and wealth creation, and its absence with unemployment and hardship, still retain much of their old potency and meaning. Overcoming these deeply entrenched views, first formulated during the years when Manchester was the 'chimney of the world', will not be an easy task.

Nor, if we reflect on the resistance to change from open coal fires, will it be easy to persuade millions of Britons to end their love affair with the car. The extraordinary affection that the nation once held for the smoky domestic hearth, the hub of Victorian social life, is rivalled by the contemporary attachment to the privately owned automobile, a technology that has come to occupy an equally dominant position in British society.[23] A car is a prized possession, affording not only social status and independence, but also the multifaceted pleasures of the driving experience:

> ... the exhilaration produced by the machine's speed and power, the sense of discovery and adventure in exploring the countryside and the larger country itself, the challenge in keeping the vehicle functional, [as well as] the satisfaction of being seen by others and oneself as the owner ... [24]

What is more, the private automobile plays a vital practical role in sustaining 'just-in-time' twenty-first century lifestyles, built around convenience and flexible mobility.[25] Relatively cheap fuel prices add to the car's considerable appeal and encourage private motoring. Despite the growing problems of congestion, accidents, and air pollution, the right to drive one's own car whenever and wherever one wants is still widely viewed as 'an inviolable individual freedom'.[26] Rolling back the excesses of Britain's car culture will require a good deal more political resolve than the state showed in the past when dealing with smoke emissions from the 'homely hearth'.

Last but not least, over recent decades there has been an upsurge of grassroots environmental activism, not just in Britain but world-wide, demanding the 'fundamental right to clean air, land, water and food'.[27] Many 'ordinary' men and women are no longer willing to tolerate the destruction of nature and the 'poisoning' of their communities without protest. At present, the efforts of environmental organisations, from mainstream institutions such as Greenpeace and Friends of the Earth, through more radical 'disorganisations' like Earth First! and Reclaim the Streets, to locally-based residents groups, are attracting strong support among British people drawn from all walks of life. At Salford, for example, the members of the Irlam and Cadishead Hazards Initiative, formed in 1990, constantly 'watch and sniff for accidental leaks' from nearby toxic waste treatment plants that, they believe, pose a grave threat to their health and safety. On detecting 'a bad odour ... the stink of bad fish or rotten eggs', local residents are urged to contact Salford's Environmental Health Department immediately to complain, thereby putting pressure on the polluting companies 'to clean up their acts'.[28] Moreover, on May Day 2000, traditionally a day to celebrate the struggles and achievements of radical working-class politics, thousands of Reclaim the Streets supporters blocked main roads in Manchester, London, and other British cities, as part of a global 'day of action' against the 'reckless pursuit of profit ... at the expense of people and ecosystems everywhere'.[29] More and more people are getting involved in the non-violent, direct actions of today's environmental movement; particularly those living in polluted and degraded urban areas. Clearly, priorities are changing where environmental issues are concerned, and growing grassroots resistance to pollution threats is helping to move us closer to a more sustainable society. The question is, will today's radical environmentalism succeed in changing government policy and cultural attitudes where nineteenth-century anti-pollution activism largely failed?

Notes

Introduction: Manchester, Air Pollution, and Urban Environmental History

1. *Guardian,* 20 January 1995, p.9.

2. Read, C. (ed.), *How Vehicle Pollution Affects Our Health,* Abacus, London, 1994.

3. For more information on anti-road protests visit the Reclaim the Streets website, http://www.gn.apc.org/rts/ and try the essays by *Aufheben* and John Jordan in McKay, G. (ed.), *DiY Culture: Party and Protest in Nineties Britain,* Verso, London, 1998.

4. Te Brake, W.H., 'Air Pollution and Fuel Crises in Preindustrial London, 1250–1650', *Technology and Culture,* Vol.16, 1975, pp.337–59.

5. Wheeler, M., 'Introduction', in *idem* (ed.), *Ruskin and Environment: The Storm-Cloud of the Nineteenth Century,* Manchester University Press, Manchester, 1995, p.3.

6. Ruskin, J., 'The Storm-Cloud of the Nineteenth Century', p.37. Re-printed in Cook, E.T. and Wedderburn, A. (eds), *The Works of John Ruskin,* Vol.XXXIV, George Allen, London, 1908.

7. For a discussion of Ruskin's views concerning air pollution see Wheeler, M., 'Environment and Apocalypse', in *idem* (ed.), *op.cit.*

8. Marcus, S., *Engels, Manchester, and the Working Class,* Weidenfeld & Nicolson, London, 1974, p.46.

9. Meadows, D.H., Meadows, D.L., and Randers, J., *Beyond the Limits: Global Collapse or a Sustainable Future,* Earthscan, London, 1995 edn, p.47.

10. Kidd, A.J., *Manchester,* Ryburn Publishing, Keele University Press, Keele, 1993, p.21.

11. *Ibid.*

12. Chaloner, W.H., 'The Birth of Modern Manchester', in Carter, C.F. (ed.), *Manchester and its Region,* Manchester University Press, Manchester, 1962, p.132. Rose, M.B., 'Introduction. The Rise of the Cotton Industry in Lancashire to 1830', in *idem* (ed.), *The Lancashire Cotton Industry: A History Since 1700,* Lancashire County Books, Preston, 1996, pp.9–10.

13. Dupree, M., 'Foreign Competition and the Interwar Period', in *ibid.*, p.265.

14. Walton, J.K., *Lancashire: A Social History, 1558–1939,* Manchester University Press, Manchester, 1990 edn, p.201.

15. Smith, R.A., *Air and Rain: The Beginnings of a Chemical Climatology,* Longmans, Green & Co., London, 1872. See also Wellburn, A., *Air Pollution and Climate Change: The Biological Impact,* Longman Scientific & Technical, Harlow, 1994, 2nd edn, p.97.

16. For a comprehensive survey, see Wohl, A.S., *Endangered Lives: Public Health in Victorian Britain,* Methuen, London, 1984 edn.

17. Flick, C., 'The Movement for Smoke Abatement in 19th-Century Britain', *Technology and Culture,* Vol.21, 1980, pp.49–50.

18. *Ibid.*

19. For example, see *ibid.*, p.38; Wohl, A.S., *op. cit.*, pp.215–16; or Newell, E., 'Atmospheric Pollution and the British Copper Industry', *Technology and Culture,* Vol.38, 1997, p.667. However, David Stradling's, *Smokestacks and Progressives: Environmentalists, Engineers, and Air Quality in America, 1881–1951,* Johns Hopkins University Press, Baltimore, 1999, does pay more attention to these important concepts than most air pollution studies.

20. Worster, D., 'History as Natural History: An Essay on Theory and Method', *Pacific Historical Review,* Vol.53, 1984, p.1.

21. See Stine, J.K. and Tarr, J.A., 'At the Intersection of Histories: Technology and the Environment', *Technology and Culture,* Vol.39, 1998, pp.601–40; Crosby, A.W., 'The Past and Present of Environmental History', *American Historical Review,* Vol.100, 1995, pp.1177–89; and White, R., 'American Environmental History: The Development of a New Historical Field', *Pacific Historical Review,* Vol.54, 1985, pp.297–335.

22. Braudel, F., *The Mediterranean and the Mediterranean World in the Age of Philip II,* Vols. I & II, Harper & Row, New York, 1972. Translation by Sîan Reynolds. For a recent discussion of French approaches to environmental history see Bess, M., Cioc, M., and Sievert, J., 'Environmental History Writing in Southern Europe', *Environmental History,* Vol.5, 2000, pp.545–6.

23. What follows is by no means an exhaustive list: Cioc, M., Linner, B-O., and Osborn, M., 'Environmental History Writing in Northern Europe', *Environmental History,* Vol.5, 2000, pp.396–406; MacKenzie, J.M., 'Empire and the ecological apocalypse: the historiography of the imperial environment', in Griffiths, T. and Robin, L. (eds), *Ecology and Empire: Environmental History of Settler Societies,* Keele University Press, Edinburgh, 1997; Coates, P., 'Clio's New Greenhouse', *History Today,* Vol.46, 1996, pp.15–22; Merricks, L., 'Environmental History', *Rural History,* Vol.7, 1996, pp.97–109; Hassan, J.A., *Prospects for Economic and Environmental History,* Manchester Metropolitan University, Occasional Paper, 1995; Williams, M., 'The relations of environmental history and historical geography', *Journal of Historical Geography,* Vol.20, 1994, pp.3–21; Sheail, J., 'Green History – The Evolving Agenda', *Rural History,* Vol.4, 1993, pp.209–23;

and Chase, M., 'Can History be Green? A Prognosis', *Rural History*, Vol.3, 1992, pp.243–51.

24. Worster, D., 'Appendix: Doing Environmental History', in *idem* (ed.), *The Ends of the Earth: Perspectives on Modern Environmental History*, Cambridge University Press, New York, 1988, p.293.

25. *Ibid.*, p.292.

26. McNeill, J.R., *Something New Under the Sun: An Environmental History of the Twentieth-century World*, Allen Lane, London, 2000, chapter 9 'More People, Bigger Cities'.

27. Melosi, M.V., *The Sanitary City: Urban Infrastructure in America from Colonial Times to the Present*, Johns Hopkins University Press, Baltimore, 2000; Tarr, J.A., *The Search for the Ultimate Sink: Urban Pollution in Historical Perspective*, University of Akron Press, Akron, 1996; Rosen, C.M. and Tarr, J.A., 'The Importance of an Urban Perspective in Environmental History', *Journal of Urban History*, Vol.20, 1994, pp.299–310; Melosi, M.V., 'The Place of the City in Environmental History', *Environmental History Review*, Vol.17, 1993, pp.1–23; Cronon, W., *Nature's Metropolis: Chicago and the Great West*, W.W. Norton, New York, 1992 edn; and Melosi, M.V. (ed.), *Pollution and Reform in American Cities, 1870–1930*, University of Texas Press, Austin, 1980. For a recent discussion of Joel Tarr's groundbreaking contribution to the field see Platt, H.L., 'The Emergence of Urban Environmental History', *Urban History*, Vol.26, 1999, pp.89–95.

28. Melosi, M.V., 'Place of the City', pp.3–5.

29. Rosen, C.M. and Tarr, J.A., *op. cit.*, p.307.

30. Worster, D., 'Doing Environmental History', p.293.

31. Cronon, W., 'A Place for Stories: Nature, History, and Narrative', *Journal of American History*, Vol.78, 1992, p.1349.

32. Nye, D.E., *Narratives and Spaces: Technology and the Construction of American Culture*, University of Exeter Press, Exeter, 1997, p.8.

33. *Ibid.*, p.9.

34. Lease, G., 'Introduction: Nature Under Fire', in Soulé, M.E. and Lease, G. (eds), *Reinventing Nature? Responses to Postmodern Deconstruction*, Island Press, Washington, 1995, p.7. See also Demeritt, D., 'Ecology, Objectivity and Critique in Writings on Nature and Human Societies', *Journal of Historical Geography*, Vol.20, 1994, pp.22–37; and Cronon, W., 'Comment: Cutting Loose or Running Aground?' in *ibid.*, pp.38–43.

35. For a wide-ranging discussion of this proposition see Cronon, W., 'Introduction: In Search of Nature', in *idem* (ed.), *Uncommon Ground: Rethinking the Human Place in Nature*, W.W. Norton, New York, 1996.

36. Demeritt, D., *op. cit.*, p.29.

37. *Ibid.*, p.3. Donald Worster, for example, argues that 'the real threat of modern historical consciousness lies … in its tendency to embrace a theory of total relativism'. In *idem*, 'Nature and the Disorder of History', in Soulé, M.E. and Lease, G. (eds), *op. cit.*, p.85n.

38. For a recent study that also foregrounds narrative constructions of environmental problems see Lichatowich, J., *Salmon Without Rivers: A History of the Pacific Salmon Crisis*, Island Press, Washington D.C., 1999.

39. Cronon, W., *Stories*, p.1372.

40. *Ibid.*

41. *Ibid.*, p.1374.

42. Again, this is not an exhaustive list: Rees, R., *King Copper: South Wales and the Copper Trade*, University of Wales Press, Cardiff, 2000; Bowler, C. and Brimblecombe, P., 'Control of Air Pollution in Manchester prior to the Public Health Act, 1875', *Environment and History*, Vol. 6, 2000, pp.71–98; Stradling, D. and Thorsheim, P., 'The Smoke of Great Cities: British and American Efforts to Control Air Pollution, 1860–1914', *Environmental History*, Vol.4, 1999, pp.6–31; Newell, E., *op. cit.*, pp.655–89; Hawes, R., 'The Control of Alkali Pollution in St. Helens, 1862–1890', *Environment and History*, Vol.1, 1995, pp.159–71; Clapp, B.W., *An Environmental History of Britain since the Industrial Revolution*, Longman, Harlow, 1994, chapters 2 and 3; Wohl, A.S., *op. cit.*, chapter 8; Dingle, A.E., '"The Monster Nuisance of All": Landowners, Alkali Manufacturers, and Air Pollution, 1828–64', *Economic History Review*, Second Series, Vol.XXXV, 1982, pp.529–548; Ashby, E. and Anderson, M., *The Politics of Clean Air*, Oxford University Press, Oxford, 1981; Flick, C., *op. cit.*, pp.29–50; Malcolm, C.V., 'Smokeless Zones – The History of Their Development', parts 1 and 2, *Clean Air*, Autumn 1976 and Spring 1977; Frankel, M., 'The Alkali Inspectorate: The Control of Industrial Air Pollution', *Social Audit*, Vol.1, 1974, pp.1–48; Macleod, R.M., 'The Alkali Acts Administration, 1863–84: The Emergence of the Civil Scientist', *Victorian Studies*, Vol.IX, 1965, pp.85–112; Beck, A., 'Some Aspects of the History of Anti-Pollution Legislation in England, 1819–1954', *Journal of the History of Medicine and Allied Sciences*, Vol.14, 1959, pp.475–489; and Graham, J., 'Smoke Abatement in Manchester: Past and Present', *Sanitarian*, 1954, pp.235–9.

43. Wohl, A., *op. cit.*, p.219; and Ashby, E. and Anderson, M., *op. cit.*

44. Flick, C., *op. cit.*

45. Porter, D., *The Thames Embankment: Environment, Technology, and Society in Victorian London*, University of Akron Press, Akron, 1998; Hassan, J.A., *A History of Water in Modern England and Wales*, Manchester University Press, Manchester, 1998; Goddard, N., '"A mine of wealth"? The Victorians and the Agricultural Value of Sewage', *Journal of Historical Geography*, Vol.22, 1996, pp.274–90; Hamlin, C., 'Environmental Sensibility in Edinburgh, 1839–1840: The "Fetid Irrigation" Controversy', *Journal of Urban History*, Vol.20, 1994, pp.311–39; Walklett, H.J., 'The Pollution of the Rivers of South-East Lancashire by Industrial Waste between c.1860 and c.1900', unpublished PhD thesis, University of Lancaster, 1993; Luckin, B., *Pollution and Control: A Social History of the Thames in the Nineteenth Century*, Adam Hilger, Bristol, 1986; Hassan, J.A., 'The Impact and Development of the Water Supply in Manchester, 1568–1882', *Transactions of the Historic Society of Lancashire and Cheshire*, Vol.133, 1984, pp.25–45; and

Richards, T., 'River Pollution Control in Industrial Lancashire, 1848–1939', unpublished PhD thesis, University of Lancaster, 1982.

46. Rosen, C.M. and Tarr, J.A., *op. cit.*, p.301.

47. Hamlin, C., *op. cit.*, p.311.

48. Douglas, M., *Purity and Danger: An Analysis of the Concepts of Pollution and Taboo*, Routledge, London, 1991 edn, p.35. See also *idem, Implicit Meanings: Essays in Anthropology*, Routledge & Kegan Paul, London, 1975, chapter 3, 'Pollution'.

49. Douglas, M. and Wildavsky, A., *Risk and Culture: An Essay on the Selection of Technical and Environmental Dangers*, University of California Press, Berkeley, 1982, p.8 and p.35.

50. Schutz, A., *On Phenomenology and Social Relations: Selected Writings*, edited by Wagner, H.R., University of Chicago Press, Chicago, 1970. See also Berleant, A., *The Aesthetics of Environment*, Temple University Press, Philadelphia, 1992.

51. Schutz, A., *op. cit.*, p.163.

52. For example, see Stradling, D. and Thorsheim, P., *op. cit.*, pp.19–22.

53. Peter Brimblecombe's book on air pollution in London is a wide-ranging and engaging piece of work, although in covering the topic from medieval times right through to the 1970s in less than two hundred pages it does lack in-depth analysis of important issues. See *idem, The Big Smoke: A History of Air Pollution in London since Medieval Times*, Routledge, London, 1988.

Part One: The Nature of Smoke

1. For a more detailed discussion of this concept, see Rosen, C.M. and Tarr, J.A., 'The Importance of an Urban Perspective in Environmental History', *Journal of Urban History*, Vol.20, 1994, pp.299–310.

2. A question that is posed in a late twentieth-century context in Eden, S., 'Public Participation in Environmental Policy: Considering Scientific, Counter-scientific, and Non-scientific Contributions', *Public Understanding of Science*, Vol.5, 1996, p.197.

3. Tuan, Y-F., *Topophilia: A Study of Environmental Perception, Attitudes, and Values*, Columbia University Press, New York, Morningside edn, 1990.

4. Milton, K., *Environmentalism and Cultural Theory: Exploring the Role of Anthropology in Environmental Discourse*, Routledge, London, 1996, p.60.

5. *Ibid.*, p.62.

6. Berleant, A., *The Aesthetics of Environment*, Temple University Press, Philadelphia, 1992, p.10 and p.132.

7. *Builder*, 12 August, 1899, p.143. See also Popplewell, W.C., *The Prevention of Smoke*, Scott, Greenwood & Co., London, 1901, p.xv.

8. *Parliamentary Papers*, (House of Commons), 1819 (574) VIII, p.5 and p.10.

9. *Parliamentary Papers*, (House of Commons), 1843 (583) VII, p.iv.

10. *Ibid.*, q.680.

11. *Manchester Guardian*, 20 September 1888.

12. Jenner, M., 'The Politics of London Air: John Evelyn's *Fumifugium* and the Restoration', *Historical Journal*, Vol.38, 1995, pp.535–51; Brimblecombe, P., *The Big Smoke: A History of Air Pollution in London since Medieval Times*, Routledge, London, 1988; and Te Brake, W.H., 'Air Pollution and Fuel Crises in Preindustrial London, 1250–1650', *Technology and Culture*, Vol.16, 1975, pp.337–59.

13. However, John Langton has described the relatively large demand for coal in the industrial hearths and foundries of seventeenth-century Wigan. See *idem, Geographical Change and Industrial Revolution: Coalmining in South West Lancashire, 1590–1799*, Cambridge University Press, Cambridge, 1979, pp.50–4.

14. Hatcher, J., *The History of the British Coal Industry, Volume 1, Before 1700: Towards the Age of Coal*, Clarendon Press, Oxford, 1993, p.118.

15. *Ibid.*, p.117.

16. *Ibid.*, pp.121–2.

17. Nef, J.U., *The Rise of the British Coal Industry*, Vol.1, George Routledge & Sons, London, 1932, p.64 and p.108.

18. Flinn, M.W., *The History of the British Coal Industry, Volume 2, 1700–1830: The Industrial Revolution*, Clarendon Press, Oxford, 1984, p.231.

19. Challinor, R., *The Lancashire and Cheshire Miners*, Frank Graham, Newcastle, p.15 and p.182.

20. Wheeler, J., *Manchester: Its Political, Social and Commercial History, Ancient and Modern*, Whittaker & Co., London, 1836, p.449.

21. For a more detailed discussion see Pollard, S., 'A New Estimate of British Coal Production', *Economic History Review*, 2nd Series, Vol.XXXIII, 1980, pp.212–35.

22. Flinn, M.W., *op. cit.*, p.449.

23. Musson, A.E., 'Industrial Motive Power in the United Kingdom, 1800–70', *Economic History Review*, 2nd Series, Vol.XXIX, 1976, p.429. See also Tunzelmann, N. von, 'Coal and Steam Power', in Langton, J. and Morris, R.J. (eds), *Atlas of Industrialising Britain, 1780–1914*, Methuen, London, 1986, pp.72–9.

24. The traditional view of Manchester as an industrial town has been downplayed in recent years, with the accent being placed instead on its expanding commercial functions. The importance of the continued growth of manufacturing industry in the city is now beginning to be reasserted. For a detailed discussion see Gunn, S., 'The Manchester Middle Class, 1850–1880', unpublished PhD thesis, University of Manchester, 1992, chapter 1. For renewed attention and prominence given to the factory more generally in Lancashire and the West Riding of Yorkshire see Gray, R., *The Factory Question and Industrial England, 1830–1860*, Cambridge University Press, Cambridge, 1996.

25. Estcourt, C., 'Why the Air of Manchester is so Impure', *Health Journal and Record of Sanitary Engineering*, Vol. IV, 1886–7, p.189.

26. Chaloner, W.H., 'The Birth of Modern Manchester', in Carter, C.F. (ed.), *Manchester and its Region*, Manchester University Press, Manchester, 1962, p.133.

27. Walton, J.K., *Lancashire: A Social History, 1558–1939*, Manchester University Press, Manchester, 1990 edn, p.107.

28. *Ibid.*

29. Musson, A.E., *op. cit.*, p.429.

30. *Ibid.*, p.435.

31. Rose, M.B., 'Introduction. The rise of the cotton industry in Lancashire to 1830', in *idem* (ed.), *The Lancashire Cotton Industry: A History Since 1700*, Lancashire County Books, Preston, 1996, p.17; Scola, R., *Feeding the Victorian City: The Food Supply of Manchester, 1770–1870*, Manchester University Press, Manchester, 1992, p.20; Kidd, A.J., *Manchester*, Ryburn Publishing, Keele University Press, Keele, 1993, p.24; and Gunn, S., *op. cit.*, Table 1.7, p.62.

32. See Mosley, S., 'The "Smoke Nuisance" and Environmental Reformers in Late Victorian Manchester', *Manchester Region History Review*, Vol. X, 1996, p.43.

33. Church, R., *The History of the British Coal Industry, Volume 3, 1830–1913: Victorian Pre-eminence*, Clarendon Press, Oxford, 1986, pp.758–9.

34. Wheeler, J., *op. cit.*, p.449n.

35. *Parliamentary Debates*, Vol.181, 1866, col.1817.

36. Church, R., *op. cit.*, p.759.

37. *Ibid.*

38. Platt, H.L., 'Invisible Gases: Smoke, Gender, and the Redefinition of Environmental Policy in Chicago, 1900–1920', *Planning Perspectives*, Vol.10, 1995, p.75.

39. See Mohun, A.P., *Steam Laundries: Gender, Technology, and Work in the United States and Great Britain*, Johns Hopkins University Press, Baltimore, 1999.

40. *Manchester Guardian*, 28 May 1842.

41. Alloway, B.J. and Ayres, D.C., *Chemical Principles of Environmental Pollution*, Blackie Academic & Professional, Glasgow, 1993, pp.199–200.

42. *Ibid.*, p.129.

43. Smith, R.A., 'On the Air of Towns', *Journal of the Chemical Society*, Vol.11, 1859, p.206; and Osborne, H., *The Problem of Atmospheric Pollution*, John Heywood, Manchester, 1924, p.23.

44. Frend, W., *Is it Impossible to Free the Atmosphere of London in a Very Considerable Degree, From the Smoke and Deleterious Vapours With Which it is Hourly Impregnated?* London, 1819, p.2.

45. Brimblecombe, P., *op. cit.*

46. Quote from Humphrey Ward, Mrs, *The History of David Grieve*, Smith, Elder, & Co., London, 6th edn, 1892, p.8. For contemporary reactions to Manchester see Bradshaw, L.D. (ed.), *Visitors to Manchester: A Selection of British and Foreign Visitors' Descriptions of Manchester from c.1538 to 1865*, Neil Richardson, Swinton,

1987; and Marcus, S., *Engels, Manchester, and the Working Class*, Weidenfeld & Nicolson, London, 1974, chapter 2, 'The Town'.

47. Cardwell, D.S.L. (ed.), *Artisan to Graduate*, Manchester University Press, Manchester, 1974, p.4.

48. Aston, J., *A Picture of Manchester*, 1816, p.223. Reprinted by E.J. Morten, Manchester, 1969.

49. Bamford, S., *Walks in South Lancashire and on its Borders*, 1844, p.9. Reprinted by Augustus M. Kelley, Clifton, New Jersey, 1972.

50. *Ibid.*, p.10.

51. Grindon, L.H., *Manchester Walks and Wild Flowers: An Introduction to the Botany and Rural Beauty of the District*, Whittaker & Co., London, 1859, pp.1–2.

52. *Ibid.*, pp.6–7.

53. Faucher, L., *Manchester in 1844; Its Present Condition and Future Prospects*, Abel Heywood, Manchester, 1844, p.16. Frank Cass, London, reprint of 1969. Major-General Sir Charles Napier is quoted in Marcus, S., *op. cit.*, p.46.

54. Ginswick, J. (ed)., *Labour and the Poor in England and Wales, 1849–1851: The Letters to the Morning Chronicle from the Correspondents in the Manufacturing and Mining Districts, the Towns of Liverpool and Birmingham, and the Rural Districts, Volume 1, Lancashire, Cheshire, Yorkshire*, Frank Cass, London, 1983, p.3.

55. Emrys-Jones, A., 'Smoke and Impure Air', *Exhibition Review*, No.4, 1882, p.3.

56. *Parliamentary Papers*, (House of Commons), 1819 (574) VIII, p.5.

57. Smith, R.A., 'Air of Towns', p.197.

58. Russell, F.A.R., *Smoke in Relation to Fogs in London*, National Smoke Abatement Institution, London, 1889, p.13.

59. Schunck, E., 'President's Address', *Journal of the Society of Chemical Industry*, Vol.XVI, 1897, p.593.

60. Moss, C.E., 'Changes in the Halifax Flora During the Last Century and a Quarter', *Naturalist*, Vol.26, 1901, pp.105–6.

61. Swindells, T., *Manchester Streets and Manchester Men*, First Series, J.E. Cornish, Manchester, 1906, p.3.

62. *Manchester Guardian*, 13 June 1888.

63. Rawnsley, H.D., 'Sunlight or Smoke?' *Contemporary Review*, Vol.57, 1890, pp.512–13.

64. Emrys-Jones, A., *op. cit.*, p.3.

65. Carpenter, E., 'The Smoke-Plague and its Remedy', *Macmillan's Magazine*, Vol.62, 1890, pp.204–5.

66. See discussion in Chubb, L.W., *Smoke Abatement*, Coal Smoke Abatement Society, London, 1912, p.12.

67. Popplewell, W.C., *op. cit.*, p.xiv.

68. Carpenter, E., *op. cit.*, p.207.

69. *Manchester Guardian*, 25 April 1902.

70. Smith, R.A., 'Some Ancient and Modern Ideas of Sanitary Economy', *Memoirs and Proceedings of the Manchester Literary and Philosophical Society*, 2nd Series, Vol.11, 1854, p.61.

71. Lowe, J., 'In Praise of Tall Chimneys', *Industrial Heritage*, Vol.8, 1989, pp.11–19.

72. Russell, F.A.R., *The Atmosphere in Relation to Human Life and Health*, Smithsonian Institution, Washington, 1896, p.3.

73. Quoted in Andersen, A., 'Die Rauchplage im deutschen Kaiserreich als Beispiel einer versuchten Umweltbewältigung', in Jaritz, G. and Winiwarter, V. (eds), *Umweltbewältigung: Die Historische Perspektive*, Verlag Für Regionalgeschichte, Bielefeld, 1994, pp.128–9. Translation by Monika Büscher. See also Brüggemeier, F-J., 'A Nature Fit for Industry: The Environmental History of the Ruhr Basin, 1840–1990', *Environmental History Review*, Vol.18, 1994, p.44. For a contemporary British discussion of the atmosphere's 'self-purifying' properties see Ransome, A., 'On Foul Air and Lung Disease', in *Health Lectures for the People*, Vol.1, Manchester and Salford Sanitary Association, Manchester, 1878, pp.41–3.

74. Spence, P., *Coal, Smoke, and Sewage, Scientifically and Practically Considered*, Cave & Sever, Manchester, 1857, pp.21–5.

75. Smith, R.A., 'Ancient and Modern', pp.82–3.

76. *Parliamentary Papers*, (House of Commons), 1819 (574) VIII, p.5. See also *Ibid.*, 1843 (583) VII, q.1003 and q.1009.

77. Langton, J., 'The physical environment', in Langton, J. and Morris, R.J. (eds), *op. cit.*, p.6.

78. Farnie, D., *The English Cotton Industry and the World Market, 1815–96*, Clarendon Press, Oxford, 1979, pp.47–51.

79. See Brimblecombe, P., *Air Composition and Chemistry*, Cambridge University Press, Cambridge, 2nd edn, 1996, pp.130–8.

80. Longhurst, J.W.S. and Conlan, D.E., 'Changing Air Quality in the Greater Manchester Conurbation', in Baldasano, J.M., Brebbia, C.A., Power, H., and Zannetti, P. (eds), *Air Pollution II Volume 1: Computer Simulation*, Computational Mechanics Publications, Southampton, 1994, p.349.

81. Simon, S.D., *A Century of City Government, Manchester 1838–1938*, George Allen & Unwin, London, 1938, p.202.

82. Smith, R.A., 'What Amendments are Required in the Legislation Necessary to Prevent the Evils arising from Noxious Vapours and Smoke?' *Transactions of the National Association for the Promotion of Social Science*, 1876, pp.513–14.

83. Davis, G.E., 'Smoke and Fogs', *Health Journal and Record of Sanitary Engineering*, Vol.1, 1883–4, p.170.

84. See 'In the Days of King Fog', *Punch*, 21 January 1888, pp.26–7; and *Daily Mail*, 13 December 1905.

85. *Manchester Guardian* 16 January 1888; and Graham, J.W., *The Destruction of Daylight: A Study in the Smoke Problem*, George Allen, London, 1907, p.25.

86. *Evening Standard and St. James Gazette*, 12 December 1905.

87. See, for example, Sims, G.R., 'London under the Weather', in *idem* (ed.), *Living London*, Vol.II, Cassell & Co., London, 1902, pp.261–5.

88. Wood, C.M., Lee, N., Luker, J.A., and Saunders, P.J.W., *The Geography of Pollution: A Study of Greater Manchester*, Manchester University Press, Manchester, 1974, p.16.

89. Niven, J. and Tattersall, C.H., 'Meteorology and Health of Manchester and Salford', in Ray, J.H. (ed.), *British Medical Association. Manchester Meeting 1902. Handbook and Guide to Manchester*, F. Ireland, Manchester, 1902, p.40; and *Report and Proceedings of the Manchester Field-Naturalists and Archaeologists' Society, For the Year 1892, Including Second Report on the Atmosphere of Manchester and Salford*, Examiner Printing Works, Manchester, 1893, p.84.

90. *Ibid.*, p.83.

91. Quoted in Davis, G.E., *op. cit.*, p.170.

92. Popplewell, W.C., 'A Smokeless London: The Promises of Gaseous Fuel and Electricity', *Cassier's Magazine*, Vol.XXI, 1902, p.211; and Brimblecombe, P., *Big Smoke*, p.112.

93. *First Annual Report of the Sanitary Committee on the Work of the Air Pollution Advisory Board*, City of Manchester, 1915, p.36; and Jenkins, W.C., 'Note on Foggy Days in Manchester', *Memoirs and Proceedings of the Manchester Literary and Philosophical Society*, Vol.59, Part 1, 1914–15, pp.1–4.

94. Russell, W.T., 'The Influence of Fog on Mortality from Respiratory Diseases', *Lancet*, 16 August 1924, p.335–8; and Niven, J. and Tattersall, C.H., *op. cit.*, p.41.

95. Chalmers, A.K. (ed.), *Public Health Administration in Glasgow: A Memorial Volume of the Writings of James Burn Russell*, James Maclehose & Sons, Glasgow, 1905, p.372.

96. Shaw, W.N., 'The Treatment of Smoke; A Sanitary Parallel', *Journal of the Sanitary Institute*, Vol.XXIII, Part III, 1902, p.324.

97. *Manchester As It Is*, Love & Barton, Manchester, 1839, p.123.

98. Philips, H., 'Open Spaces for Recreation in Manchester', *Transactions of the Manchester Statistical Society*, 1896–7, p.50.

99. *Parliamentary Papers*, (House of Commons), 1843 (583) VII, q.869 and q.871.

100. Wood, C.M., Lee, N., Luker, J.A., and Saunders, P.J.W., *op. cit.*, p.15.

101. *Ibid.*

102. Schunck, E., *op. cit.*, p.593.

103. Cowling, E.B., 'Acid Precipitation in Historical Perspective', *Environmental Science and Technology*, Vol.16, 1982, p.111A.

104. Cohen, J.B. and Ruston, A.G., *Smoke: A Study of Town Air*, Edward Arnold, London, 1912, pp.70–1.

105. For example, see Battarbee, R.W., 'Diatom Analysis and the Acidification of Lakes', *Philosophical Transactions of the Royal Society of London*, Series B, Vol.305, 1984, pp.451–77.

106. Irwin, W., 'The Soot Deposited on Manchester Snow', *Journal of the Society of Chemical Industry*, Vol.XXI, Part 1, 1902, p.533.

107. See *First Annual Report on Air Pollution Advisory Board*, pp.13–18; and *Times Engineering Supplement*, 27 November 1914.

108. For the conversion of the values used by the Air Pollution Advisory Board in 1915 I am indebted to Peter Lucas of the Institute of Environmental and Biological Sciences, University of Lancaster.

109. Brimblecombe, P., *Air Composition*, pp.93–4; and Wellburn, A., *Air Pollution and Climate Change: The Biological Impact*, Longman Scientific & Technical, Harlow, 2nd edn, 1994, chapter 2.

110. Smith, R.A., 'What Amendments', p.516.

111. Schwela, D., 'Vergleich der nassen Deposition von Luftverunreinigungen in den Jahren um 1870 mit heutigen Belastungswerten', *Staub-Reinhalt. Luft*, Vol.43, 1983, pp.135–9. Translation by Monika Büscher. Although the reliability of Smith's scientific work has been questioned. See Gibson, A. and Farrar, W.V., 'Robert Angus Smith, F.R.S., and "Sanitary Science"', *Notes and Records of the Royal Society of London*, Vol.28, 1973–4, p.253.

112. Shaw, W.N., *op. cit.*, p.323.

113. For electrically-driven fans see *ibid.*, p.328; for electrical discharges see Russell, F.A.R., *Smoke*, p.36 and Des Voeux, H.A., 'How to Abate the Smoke Nuisance', *Church Family Newspaper*, 20 January 1905; for water sprinklers see Schunck, E., *op. cit.*, p.593.

114. Smith, R.A., 'What Amendments', p.514.

115. Defoe, Daniel, *A Tour Through the Whole Island of Great Britain*, Vol.2, J.M. Dent & Sons, London, 1962, p. 260. First published 1724–6.

116. Grindon, L.H., *Walks*, p.2.

117. Casartelli, L.C., 'On Town Beauty', *Transactions of the Manchester Statistical Society*, 1898–9, p.8.

118. Grindon, L.H., *Walks*, pp.2–3.

119. Casartelli, L.C., *op. cit.*, p.8.

120. Engels, F., *The Condition of the Working Class in England*, Penguin, Harmondsworth, 1987 edn, p.92. First published in Germany in 1845.

121. *Manchester Guardian*, 28 May 1842.

122. Philips, H., *op. cit.*, pp.50–1.

123. Hamilton, E., 'Smoke', *Pall Mall Magazine*, Vol.II, 1894, p.401.

124. Roberts, R., *The Classic Slum: Salford Life in the First Quarter of the Century*, Penguin, Harmondsworth, 1990 edn, p.16 and pp.238–9.

125. Philips, H., *op. cit.*, pp.51–3.

126. *Bradshaw's Illustrated Guide to Manchester*, Bradshaw & Blacklock, Manchester, 1857, p.20. However, by the 1880s it was argued that Manchester did not measure up to standards set by other British industrial cities with regard to its provision of

open spaces. After the opening of its first public parks, Manchester, it was alleged, had remained practically stationary in the matter of providing parks and open spaces for its citizens. According to Herbert Philips, in 1896 Manchester had 312 acres and Salford just 122 acres of parks and playgrounds as against: 850 acres at Liverpool; over 817 acres at Glasgow; and some 780 acres of parkland at Leeds. Not until 1903, when the city purchased Heaton Park, comprising 650 acres, was progress made towards meeting the needs of a growing population for more space for outdoor recreation. See: Philips, H., *op. cit.*, pp.49–64.

127. Bradford Local Board Report. Press cutting in Salford Local History Library, ref. 614.71.

128. Philips, H., *op. cit.*, p.59.

129. Conway, H., *People's Parks: The Design and Development of Victorian Parks in Britain*, Cambridge University Press, Cambridge, 1991, p.180. A similar selection process was taking place in industrialising Germany by the late 1870s. See Schramm, E., 'Experts in the Smelter Smoke Debate', in Brimblecombe, P. and Pfister, C. (eds), *The Silent Countdown: Essays in European Environmental History*, Springer-Verlag, Berlin/Heidelberg, 1990, pp.199–200; and Brüggemeier, F-J., *op. cit.*, p.42.

130. Haweis, Mrs, *Rus in Urbe: or Flowers that Thrive in London Gardens and Smoky Towns*, Leadenhall Press, London, 1886, p.22.

131. Thomas, K., *Man and the Natural World: Changing Attitudes in England 1500–1800*, Penguin, Harmondsworth, 1984 edn, p.233.

132. Grindon, L.H., 'A Naturalist on Smoke', *Exhibition Review*, No.5, 1882, p.1.

133. *Ibid.*

134. *Manchester Guardian*, 16 August 1887.

135. Pettigrew, W.W., *The Influence of Air Pollution on Vegetation*, Smoke Abatement League of Great Britain, Pamphlet, 1928, p.4.

136. *Ibid*, P.5; and Birkbeck, W., 'Manchester Parks and Recreation Grounds', in Sutton, C.W. (ed.), *Handbook and Guide to Manchester*, Sherratt & Hughes, Manchester, 1908, p.73.

137. *Ibid*, p.74.

138. Pettigrew, W.W., *op. cit.*, p.6.

139. Cohen, J.B. and Ruston, A.G., *Smoke*, p.27.

140. Quoted in Rolleston, C., 'The Cloud over English Life', *Westminster Review*, Vol.162, 1904, p.34.

141. *Parliamentary Papers*, (House of Commons), 1843 (583) VII, q.1007.

142. Grindon, L.H., *Walks*, p.4.

143. Secord, A., 'Science in the Pub: Artisan Botanists in Early Nineteenth-century Lancashire', *History of Science*, Vol.32, 1994, pp.287–8.

144. Grindon, L.H., *Walks*, p.142.

145. Grindon, L.H., *The Manchester Flora: A Descriptive List of the Plants Growing Wild Within Eighteen Miles of Manchester*, William White, London, 1859, p.513.

146. *Ibid.*, p.511.

147. See Richardson, D.H.S., *Pollution Monitoring with Lichens*, Richmond Publishing, Slough, 1992.

148. Ferguson, P. and Lee, J.A., 'Past and Present Sulphur Pollution in the Southern Pennines', *Atmospheric Environment*, Vol.17, 1983, p.1131.

149. Ratcliffe, D.A. (ed.), *A Nature Conservation Review*, Vol.1, Cambridge University Press, Cambridge, 1977, p.315.

150. Emrys-Jones, A., *op. cit.*, p.3. Also see the comments regarding south-east Lancashire's vegetation of the noted Manchester social reformer Charles Rowley in his autobiography *Fifty Years of Work Without Wages*, Hodder & Stoughton, London, 2nd edn, n.d., p.4. First published in 1912.

151. Penn, M., *Manchester Fourteen Miles*, Futura, London, 1982 edn. First published in 1947.

152. *Report of the Manchester Field-Naturalists and Archaeologists' Society for 1892*, p.82.

153. Philips, H., *op. cit.*, p.54. However, there is evidence to suggest that the indigenous ryegrass of the Manchester region had developed a tolerance to sulphur dioxide by the 1970s. See Horsman, D.C., Roberts, T.M., and Bradshaw, A.D., 'Evolution of Sulphur Dioxide Tolerance in Perennial Ryegrass', *Nature*, Vol.276, 1978, p.493.

154. *Bradshaw's Guide*, p.5.

155. *Black's Guide to Manchester*, A. & C. Black, London, 1913, pp.i–iv.

156. Chorley, K., *Manchester Made Them*, Faber & Faber, London, 1950, p.41.

157. Glick, T.F., 'Science, Technology, and the Urban Environment: The Great Stink of 1858', in Bilsky, L.J. (ed.), *Historical Ecology: Essays on Environment and Social Change*, Kennikat Press, Port Washington, New York, 1980, p.126.

158. Dickens, C., 'Smoke or No Smoke', *Household Words*, Vol.9, 1854, p.465; and Cohen, J.B. and Ruston, A.G., 'The Nature, Distribution and Effects upon Vegetation of Atmospheric Impurities in and near an Industrial Town', *Journal of Agricultural Science*, Vol.4, 1911, p.53. However, for more evidence of animals suffering due to severe air pollution see 'The Smithfield Club Cattle Show', *Illustrated London News*, 20 December 1873, p.611; and Rees, R., *King Copper: South Wales and the Copper Trade*, University of Wales Press, Cardiff, 2000, pp.79–81 and pp.117–20.

159. Brend, W.A., *Health and the State*, Constable & Co., London, 1917, p.149.

160. Saleeby, C.W., *Sunlight and Health*, Nisbet & Co., London, 5th edn, 1929, chapter XI. First published in 1923.

161. Mitchell, F.S., *The Birds of Lancashire*, Gurney & Jackson, London, 2nd edn, 1892, p.ix. First published in 1885.

162. *Ibid.*, pp.ix –x.

163. Oakes, C., *The Birds of Lancashire*, Oliver & Boyd, Edinburgh, 1953, p.185.

164. Cohen, J.B., *The Character and Extent of Air Pollution in Leeds*, Goodall & Suddick, Leeds, 1896, pp.13–14.

165. Barnett, Mrs S.A., 'The Children's Country Holiday Fun', in Barnett, Canon S.A. and Barnett, Mrs S.A., *Practicable Socialism*, Longmans, Green & Co., London, 1915, p.47.

166. Cherfas, J., 'Clean Air Revives the Peppered Moth', *New Scientist*, Vol.109, 1986, p.17; and Cook, L.M., Askew, R.R., and Bishop, J.A., 'Increasing Frequency of the Typical Form of the Peppered Moth in Manchester', *Nature*, Vol.227, 1970, p.1155.

167. Kidd, A.J., *Manchester*, p.106; and Stewart, C., *The Stones of Manchester*, Edward Arnold, London, 1956, p.36.

168. Stewart, C., 'The Battlefield: A Pictorial Review of Victorian Manchester', *Journal of the Royal Institute of British Architects*, Vol.67, 1960, p.238.

169. Disraeli, B., *Coningsby: or the New Generation*, John Lehmann, London, 1948 edn, p.152. First published in 1844. And *Bradshaw's Guide*, p.18.

170. Brooks, M.W., *John Ruskin and Victorian Architecture*, Thames & Hudson, London, 1989, p.227.

171. Popplewell, W.C., *Prevention*, p.xiii; Casartelli, L.C., *op. cit.*, p.16; and Brend, W.A., *op. cit.*, pp.92–3.

172. However, for a cautious appraisal of the practical attributes of glazed surfaces see Stratton, M., *The Terracotta Revival: Building Innovation and the Image of the Industrial City in Britain and North America*, Victor Gollancz, London, 1993, p.17.

173. Hitchcock, H-R., 'Victorian Monuments of Commerce', *Architectural Review*, Vol.105, 1949, p.64; and Stewart, C., *Stones*, p.37.

174. See Brooks, M.W., *op. cit.*, p.226; and Murgatroyd, J., 'An Architect's View of the Smoke Question', *Exhibition Review*, No.4, 1882, p.2.

175. Wilmer, C., 'Introduction', in *idem* (ed.), *Unto This Last, and Other Writings by John Ruskin*, Penguin, Harmondsworth, 1985, p.15.

176. Cook, E.T. and Wedderburn, A. (eds), *The Works of John Ruskin, Vol.IX, The Stones of Venice Vol.1, The Foundations*, George Allen, London, 1903, p.411. First published in 1851.

177. Wilmer, C., *op. cit.*, p.15.

178. Brooks, M.W., *op. cit.*, p.175.

179. *The Inauguration of the Albert Memorial, Manchester*, R. Clark, Edinburgh, 1867, pp.10–11.

180. See Thompson, P., 'The Building of the Year: Manchester Town Hall', *Victorian Studies*, Vol.11, 1967–8, pp.401–3.

181. Pass, A., 'Thomas Worthington: Practical Idealist', *Architectural Review*, Vol.155, 1974, p.268.

182. Smith, R.A., 'Air of Towns', p.232.

183. For recent discussions of this theme see the essays by Kate Hill and Simon Gunn in Kidd, A.J. and Nicholls, D. (eds), *Gender, Civic Culture and Consumerism: Middle-Class Identity in Britain, 1800–1940*, Manchester University Press, Manchester, 1999.

184. *Builder*, 29 January 1859, p.75.

185. Horsfall, T.C., 'The Government of Manchester', *Transactions of the Manchester Statistical Society*, 1895–6, pp.6–7. See also Kidd, A.J., 'The Industrial City and its Pre-industrial Past: the Manchester Royal Jubilee Exhibition of 1887', *Transactions of the Lancashire and Cheshire Antiquarian Society*, Vol.89, 1993, pp.72–3.

186. *Manchester Guardian*, 15 August 1887.

187. *Manchester Guardian*, 9 November 1889.

188. Horsfall, T.C., *op. cit.*, p.12.

189. Thomson, W., 'The Smoke-Polluted Atmosphere of Manchester', in *Annual Report of the Manchester and Salford Sanitary Association for 1914*, Sherratt & Hughes, Manchester, 1915, p.30; and Ashworth, J.R., *Smoke and the Atmosphere: Studies from a Factory Town*, Manchester University Press, Manchester, 1933, p.60.

190. Thomson, W., *op. cit.*, p.30.

191. *Report of the Manchester Field-Naturalists and Archaeologists' Society for 1892*, pp.92–3; and Ashworth, J.R., *op. cit.*, p.104.

192. *Public Health Engineer*, Vol.XVII, 1905, p.127 and p.141.

193. Roberts, E., *A Woman's Place: An Oral History of Working-Class Women 1890–1940*, Basil Blackwell, London, 1985 edn, pp.125–8.

194. Ashworth, J.R., *op. cit.*, p.71; and Davidson, C., *A Woman's Work is Never Done: A History of Housework in the British Isles 1650–1950*, Chatto & Windus, London, 1982, pp.152.

195. For example, see Davies, C.S., *North Country Bred: A Working-Class Family Chronicle*, Routledge & Kegan Paul, London, 1963, p.108.

196. Ashworth, J.R., *op. cit.*, p.61.

197. Dupree, M., 'Foreign Competition and the Interwar Period', in Rose, M.B. (ed.), *op. cit.*, pp.265–6; and Walton, J.K., *op. cit.*, p.326.

198. Richardson, C.J., *The Englishman's House from a Cottage to a Mansion*, John Camden Hotten, London, 1870, p.404.

199. *Times*, 2 January 1855.

200. Hamilton, E., *op. cit.*, p.399.

201. *Report of the Manchester Field-Naturalists and Archaeologists' Society for 1892*, p.93.

202. *Ibid.*, pp.90–3. Moreover, in 1913 Davyhulme sewage works, situated five miles south-west of Manchester, received over twice as much sunlight as the city centre's School of Technology. See Shaw, Sir N. and Owens, J.S., *The Smoke Problem of Great Cities*, Constable & Co., London, 1925, p.64.

203. Nicholson, W., *Smoke Abatement: A Manual for the Use of Manufacturers, Inspectors, Medical Officers of Health, Engineers, and Others*, Charles Griffin, London, 2nd edn, 1927, pp.50–2. First published in 1905.

204. Brodie, F.J., 'The Incidence of Bright Sunshine over the United Kingdom During the Thirty Years 1881–1910', *Quarterly Journal of the Royal Meteorological Society*, Vol.XLII, 1916, pp.34–5.

205. Beeton, Mrs I., *Book of Household Management*, Ward, Lock & Co., London, 1906 edn, p.1776. First published in 1861. See also Greenwood, W., *Love on the Dole: A Tale of the Two Cities*, Jonathan Cape, London, 1935 edn, pp.16–17. First published in 1933.

206. Horsfield, M., *Biting the Dust: The Joys of Housework*, Fourth Estate, London, 1997, pp.60–1.

207. Although not exclusively, see Brooks, J.B., *Lancashire Bred: An Autobiography*, Church Army Press, Oxford, n.d., pp.37–8. First published in two parts, 1950–1.

208. Walton, J.K., *op. cit.*, p.287.

209. For husbands helping with household chores, see Roberts, R., *op. cit.*, p.54.

210. Bamford, S., *op. cit.*, pp.32–3 and pp.68–9; and Reach, A.B., *Manchester and the Textile Districts in 1849*, Helmshore Local History Society, 1972 reprint, p.34 and p.85.

211. G., J.M., 'The Smoke Nuisance: A Lady's View', *Exhibition Review*, No.3, 1882, p.2.

212. Reeves, M.P., *Round About a Pound a Week*, Virago Press, London, 1979, pp.60–1. First published in 1913.

213. Roberts, R., *op. cit.*, p.37.

214. Russell, F.A.R., *Smoke*, p.21.

215. For example, see Oats, H.C., 'Inquiry into the Educational and other Conditions of a District in Ancoats', *Transactions of the Manchester Statistical Society*, 1865–66, pp.1–16.

216. Wilkinson, T.W., 'Van Dwelling London', in Sims, G.R. (ed.), *Living London*, Vol.III, Cassell & Co., London, 1903, p.323; and Ormerod, F., *Lancashire Life and Character*, Edwards & Bryning, 2nd edn, 1915, p.54.

217. Brierley, B., *'Ab-o'th'-Yate' Sketches and other Short Stories*, Vol.III, W.E.Clegg, Oldham, 1896, p.89.

218. Tuan, Y-F., *op. cit.*, pp.65–6.

219. Cohen, J.B., *Character*, p.6. For smoke and fog driving people indoors see also Hill, L. and Campbell, A., *Health and Environment*, Edward Arnold, London, 1925, pp.3–4.

220. Gaskell, E., *Mary Barton*, Oxford University Press, Oxford, 1987 edn, pp.12–14. First published in 1848.

221. *Ibid.*, p.13.

222. Praz, M., *An Illustrated History of Interior Decoration from Pompeii to Art Nouveau*, Thames & Hudson, London, 1964, pp.64–5; and Ormerod, F., *op. cit.*, pp.55–6.

223. Roberts, R., *op. cit.*, p.34 n. See also the wonderful illustration of a working-class mantelpiece in Godwin, G., *Town Swamps and Social Bridges*, Routledge, Warnes,

& Routledge, London, 1859, p.19. Reprinted by Leicester University Press, New York, 1972.

224. Eastlake, C.L., *Hints on Household Taste in Furniture, Upholstery and Other Details*, Longmans, Green, & Co., London, 3rd edn, 1872, p.51. Reprinted by Gregg International Publishers, Farnborough, 1971.

225. *Ibid.*, pp.93–4; Ormerod, F., *op. cit.*, p.55; and Dutton, R., *The Victorian Home: Some Aspects of Nineteenth-century Taste and Manners*, B.T. Batsford, London, 1954.

226. For example, see Praz, M., *op. cit.*, p.65; Dutton, R., *op. cit.*, pp.147–8; and Gloag, J., *The Englishman's Castle: A History of Houses, Large and Small, in Town and Country, from A.D. 100 to the Present Day*, Eyre & Spottiswoode, London, 1944, pp.144–8.

227. Air Pollution Advisory Board of the Manchester City Council, *The Black Smoke Tax: An Account of Damage Done by Smoke, with an Inquiry into the Comparative Cost of Family Washing in Manchester and Harrogate*, Henry Blacklock & Co., Manchester, 1920, p.12.

228. Dutton, R., *op. cit.*, pp.20–2 and p.42; Long, H., *The Edwardian House: the Middle-Class Home in Britain 1880–1914*, Manchester University Press, Manchester, 1993, p.155; and Horsfield, M., *op. cit.*, p.123.

229. Air Pollution Advisory Board, *op. cit.*, p.4.

230. Gloag, J., *op. cit.*, p.149.

231. Leigh, J., 'Coal Smoke: Report to the Health and Nuisance Committees of the Corporation of Manchester', *Health Journal and Record of Sanitary Engineering*, Vol.1, 1883–4, p.53; and Hamilton, H., *Scientific Treatise on Smoke Abatement*, Sherratt & Hughes, Manchester, 1917, p.95.

232. Quoted in Marsh, A., *Smoke: The Problem of Coal and the Atmosphere*, Faber & Faber, London, 1947, p.113.

233. Air Pollution Advisory Board, *op. cit.*, p.6.

234. Penn., M., *op. cit.*, p.202.

235. See, for example, the advertisement for Smalley's Bronchial Essence or Universal Family Cough Elixer in Baron, W., *Bill-o'-Jack's Lancashire Monthly*, No.26, 1911, p.16. That for Owbridge's Lung Tonic can be found in Hartley, J., *Original Clock Almanac for 1905*, W. Nicholson & Sons, Wakefield, 1905, inner cover.

236. Ross, E., *Love and Toil: Motherhood in Outcast London, 1870–1918*, Oxford University Press, New York, 1993, p.166.

237. For a recent discussion and bibliography, see Hamlin, C., 'State Medicine in Great Britain', in Porter, D. (ed.), *The History of Public Health and the Modern State*, Clio Medica 26/The Wellcome Institute Series in the History of Medicine, Rodopi B.V., Amsterdam, 1994, pp.132–64.

238. Szreter, S., 'The Importance of Social Intervention in Britain's Mortality Decline c.1850–1914: a Re-interpretation of the Role of Public Health', *Social History of Medicine*, Vol.1, 1988, p.16.

239. There are some problems with the data, however, as doctors often confused bronchitis and respiratory tuberculosis fatalities. Even so, taking the two categories together, there was still little overall decrease in mortality from respiratory diseases before 1900. *Ibid.*, p.13 and p.16.

240. Wohl, A.S., *Endangered Lives: Public Health in Victorian Britain*, Methuen, London, 1984 edn, pp.10–11.

241. Pooley, M.E. and Pooley, C.G., 'Health, Society and Environment in Victorian Manchester', in Woods, R. and Woodward, J. (eds), *Urban Disease and Mortality in Nineteenth-century England*, Batsford Academic & Educational, London, 1984, p.156.

242. Kidd, A.J., *Manchester*, p.126.

243. Tatham, J., *Thirteenth Annual Report on the Health of Salford, 1881*, J. Roberts, Salford, 1882, p.12. See also Graham, J.W., *Destruction of Daylight*, p.3.

244. *Annual Report of the Manchester and Salford Sanitary Association for 1876*, Powlson & Sons, Manchester, 1877, p.9.

245. Ransome, A., 'Foul Air', p.63.

246. Thresh, J.C., *An Enquiry into the Causes of the Excessive Mortality in No.1 District, Ancoats*, John Heywood, Manchester, 1889, pp.3–4.

247. *Ibid.*, Appendix: 'The True Death-Rate of No.1 Sanitary District, Ancoats'.

248. Cohen, J.B. and Ruston, A.G., *Smoke*, Appendix A, p.73.

249. Russell, W.T., *op. cit.*, p.336.

250. Russell, F.A.R., *Atmosphere*, pp.32–3; and *idem, London Fogs*, Edward Stanford, London, 1880, pp.26–8.

251. Graham, J.W., 'Coal Smoke: Its Causes, Consequences, and Cures', *Transactions of the Manchester Statistical Society*, 1913–14, p.46. See also: Niven, J. and Tattersall, C.H., *op. cit.*, p.41; and Osborne, H., *op. cit.*, pp.7–9.

252. Russell, W.T., *op. cit.*, p.338.

253. Hardy, A., 'Rickets and the Rest: Child-care, Diet and the Infectious Children's Diseases, 1850–1914', *Social History of Medicine*, Vol.5., 1992, pp.392–3. See also Brend, W.A., *op. cit.*, pp.87–93.

254. Hardy, A., 'Rickets', p.390; and Wohl, A.S., *op. cit.*, chapter 2.

255. Beeton, Mrs I., *op. cit.*, pp.1848–9; and *Cassell's Household Guide to Every Department of Practical Life*, Vol.I, Cassell Petter & Galpin, London, n.d., p.115. First published 1869–71.

256. Hardy, A., *The Epidemic Streets: Infectious Disease and the Rise of Preventive Medicine, 1856–1900*, Clarendon Press, Oxford, 1993, p.215; and Tatham, J., 'Dust Producing Occupations', in Oliver, T. (ed.), *Dangerous Trades: The Historical, Social, and Legal Aspects of Industrial Occupations as Affecting Health, By a Number of Experts*, John Murray, London, 1902, pp.134–65.

257. Leigh, J., *op. cit.*, pp.52–3.

258. Tatham, J., *Health of Salford*, p.11.

259. See Hardy, A., *Epidemic*, p.233 and p.244.

260. Ransome, A., 'The Smoke Nuisance. (A Sanitarian's View)', *Exhibition Review*, No.1, 1882, p.3.

261. See Russell, J.B., 'The House', 1881, in Chalmers, A.K., *op. cit.*, pp.164–5; and *Manchester Guardian*, 25 April 1902.

262. *Ibid.*

263. Saleeby, C.W., *op. cit.*, p.14; and Hardy, A., *Epidemic*, p.259.

264. *Manchester Guardian*, 25 April 1902.

265. Hardy, A., 'Rickets', pp.397–9; and Loomis, W.F., 'Rickets', *Scientific American*, Vol.223, 1970, pp.76–91.

266. Owen, I., 'Reports of the Collective Investigation Committee of the British Medical Association. Geographic Distribution of Rickets, Acute and Subacute Rheumatism, Chorea, Cancer, and Urinary Calculus. In the British Islands', *British Medical Journal*, 19 January 1889, pp.113–16.

267. *Lucretia*, 'Smoke, and its Effects on Health', *Exhibition Review*, No.5, 1882, p.2.

268. Hardy, A., 'Rickets', p.391.

269. Taylor, J.S., *Smoke and Health*, Pontefract Brothers, Manchester, 1929, p.9.

270. *Times*, 2 January 1855.

271. Emrys-Jones, A., *op. cit.*, p.3.

272. Wallin, J.E.W., *Psychological Aspects of the Problem of Atmospheric Smoke Pollution*, University of Pittsburgh, Pittsburgh, 1913, pp.7–9 and pp.41–3.

273. Graham, J.W., 'Coal Smoke', p.40.

274. Berleant, A., *op. cit.*, p.10.

Part Two: Stories about Smoke

1. Merleau-Ponty, M., *Phenomenology of Perception*, Routledge, London, 1996 edn, pp.3–5. Translation by Colin Smith.

2. Berleant, A., *The Aesthetics of Environment*, Temple University Press, Philadelphia, 1992; and Bourassa, S.C., *The Aesthetics of Landscape*, Belhaven Press, London, 1991, ch.2.

3. Greider, T. and Garkovich, L., 'Landscapes: The Social Construction of Nature and the Environment', *Rural Sociology*, Vol.59, 1994, pp.1–24.

4. Cronon, W., 'A Place for Stories: Nature, History, and Narrative', *Journal of American History*, Vol.78, 1992, p.1369. Maarten Hajer's recent study of environmental discourse in Great Britain and the Netherlands was also influential in the writing of this section of the book. See Hajer, M.A., *The Politics of Environmental Discourse: Ecological Modernisation and the Policy Process*, Clarendon Press, Oxford, 1995.

5. Horsfall, T.C., 'The Government of Manchester', *Transactions of the Manchester Statistical Society*, 1895–96, p.19.

6. Procter, R.W., *Memorials of Manchester Streets,* Thos. Sutcliffe, Manchester, 1874, pp.40–2.

7. Fyfe, P., 'The Pollution of the Air: Its Causes, Effects, and Cure', in Smoke Abatement League of Great Britain, *Lectures Delivered in the Technical College, Glasgow 1910–1911,* Corporation of Glasgow, Glasgow, 1912, p.77. See also 'Lancashire' in Clarke, A., *Moorlands and Memories, Rambles and Rides in the Fair Places of Steam-Engine Land,* Tillotsons, Bolton, 2nd edn, 1920, p.9; Jackson, H., 'The Mill has Shut Down', *Cotton Factory Times,* 27 March 1885; and 'Song of the Copper Smoke' in Newell, E., 'Atmospheric Pollution and the British Copper Industry, 1690–1920', *Technology and Culture,* Vol.38, 1997, p.667.

8. Walton, J.K., *Lancashire: A Social History, 1558–1939,* Manchester University Press, Manchester, 1990 edn, p.202.

9. *Ibid.,* pp.200–1. See also Andrew Marrison's chapter in Rose, M.B. (ed.), *The Lancashire Cotton Industry: A History Since 1700,* Lancashire County Books, Preston, 1996, especially pp.246–53.

10. Taylor, W.C., *Notes of a Tour in the Manufacturing Districts of Lancashire,* Duncan & Malcolm, London, 1842, p.22. Frank Cass, London, reprint of 1968.

11. Harland, J. (ed.), *Lancashire Lyrics: Modern Songs and Ballads of the County Palatine,* Whittaker & Co., London, 1866, pp.289–92.

12. See also Waugh, E., *Lancashire Sketches,* Second Series, John Heywood, Manchester, 1892, p.182.

13. Ginswick, J. (ed.), *Labour and the Poor in England and Wales, 1849–1851: The Letters to the Morning Chronicle from the Correspondents in the Manufacturing and Mining Districts, the Towns of Liverpool and Birmingham, and the Rural Districts, Volume 1, Lancashire, Cheshire, Yorkshire,* Frank Cass, London, 1983, p.5.

14. *Parliamentary Debates,* Vol.CVII, 1849, col.206.

15. *Manchester Guardian,* 20 June 1891.

16. Nicholson, W., *Smoke Abatement: A Manual for the use of Manufacturers, Inspectors, Medical Officers of Health, Engineers, and Others,* Charles Griffin, London, 2nd edn, 1927, pp.63–4. First published in 1905.

17. Wright, A.R., 'Cooking by Gas', *Exhibition Review,* No.5, 1882, p.3.

18. Roberts, R., *A Ragged Schooling: Growing up in the Classic Slum,* Manchester University Press, Manchester, 1976, p.73. See also Mayhew, H., *London Labour and the London Poor; A Cyclopædia of the Condition and Earnings of Those that Will Work, Those that Cannot Work, and Those that Will not Work,* Vol.2, Dover Publications, New York, 1968, pp.85–7. First published in 1861. And Dickens, C., *Bleak House,* Bantam Books, New York, 1983 edn, ch.10, 'The Law-Writer', p.127. First published in 1852–3.

19. Waugh, E., *Poems and Songs,* John Heywood, Manchester, 1893, pp.107–9. See also 'Manchester Song' in Gaskell, E., *Mary Barton: A Tale of Manchester Life,* Oxford University Press, Oxford, 1987 edn, p.63. First published in 1848.

20. Smith, R.A., 'What Amendments are Required in the Legislation Necessary to Prevent the Evils Arising from Noxious Vapours and Smoke?' *Transactions of the National Association for the Promotion of Social Science*, 1876, p.513.

21. Ravetz, A., with Turkington, R., *The Place of Home: English Domestic Environments, 1914–2000*, E. & F.N. Spon, London, 1995, p.124.

22. See the list of fireplace mottoes in American Face Brick Association, *The Home Fires: A Few Suggestions in Face Brick Fireplaces*, Rogers & Co., Chicago, 1923, pp.24–5.

23. Dickens, C., *Christmas Books*, T.Nelson & Sons, London, n.d., pp.88–91. First published in 1845.

24. Brierley, B., '*Ab – O'th' – Yate' Sketches and other Short Stories*, Vol.III, W.E.Clegg, Oldham, 1896, pp.239–41; and *idem, Tales and Sketches of Lancashire Life*, Abel Heywood & Son, Manchester, 1885, p.197. See also Dickens, C., *A Christmas Carol: A Ghost Story of Christmas*, Penguin, Harmondsworth, 1984 edn, p.83. First published in 1843.

25. See, for example, 'The Wind Storm' and other short stories in Waugh, E., *The Chimney Corner*, John Heywood, Manchester, 1892. First published in 1874.

26. Budgen, L.M., *Live Coals; or, Faces from the Fire*, L.Reeve & Co., London, 1867, p.27; and Trevor, W., *Play-Times in a Busy Life: Sketches and Meditations by William Trevor*, W.E.Clegg, Oldham, 1909, p.79.

27. Budgen, L.M., *op. cit.*, p.27.

28. American Face Brick Association, *op. cit.*, p.5.

29. Waugh, E., *Poems and Songs*, pp.34–5. See also Baron, W., 'At Hooam' and 'A Seeat at Yor Own Fireside', in *idem, Bits O' Broad Lancashire, Poems in the Dialect*, John Heywood, Manchester, c.1888, pp.34–6 and pp.43–5.

30. Bone, W.A., *Coal and Health*, Pamphlet, London, 1919, p.11.

31. See Bourassa, S.C., *op. cit.*, pp.22–3.

32. Corbin, A., *The Foul and the Fragrant: Odour and the French Social Imagination*, Picador, London, 1994 edn, p.6. Translation by Miriam L. Kochan, Roy Porter, and Christopher Prendergast.

33. There was a very strong reliance placed on the sense of smell in relation to complaints about air pollution dealt with by the Manchester Corporation Nuisance Committee, particularly in the 1850s and 1860s. Unfortunately, the minutes of the committee are incomplete due to fire damage. See *Draft Minutes of the Manchester Corporation Nuisance Committee*, 1856–61 and 1880–84. Manchester Central Library Ref., M.9/65/1/1–2 and 8–11.

34. Corbin, A., *op. cit.*, p.4.

35. See Hannaway, C., 'Environment and Miasmata' in Bynum, W.F. and Porter, R. (eds), *Companion Encyclopaedia of the History of Medicine*, Volume 1, Routledge, London, 1993, ch.15; and Wohl, A.S., *Endangered Lives: Public Health in Victorian Britain*, Methuen, London, 1984 edn, pp.87–91.

36. Porter, R., 'Foreward' in Corbin, A., *op. cit.*, p.vi.

37. For example, see Chadwick, E., *Report on the Sanitary Condition of the Labouring Population of Great Britain*, 1842, Flinn, M.W. (ed.), Edinburgh University Press, Edinburgh, 1965; and Chadwick, E., *A Supplementary Report on the Results of a Special Inquiry into the Practice of Internment in Towns*, W. Clowes & Sons, London, 1843.

38. For an excellent new interpretation of Chadwick's work see Hamlin, C., *Public Health and Social Justice in the Age of Chadwick: Britain, 1800–1854*, Cambridge University Press, Cambridge, 1998.

39. Dickens, C., *Bleak House*, ch.46, 'Stop Him!' pp.582–3: see also ch.22, 'Mr Bucket', pp.288–9 and ch.32, 'The Appointed Time', pp.413–4.

40. Gaskell, E., *Mary Barton*, p.68.

41. Smith, R.A., 'Some Ancient and Modern Ideas of Sanitary Economy', *Memoirs of the Literary and Philosophical Society of Manchester*, 2nd Series, Vol.11, 1854, pp.47–9; and Corbin, A., *op. cit.*, p.64 and p.97.

42. Spence, P., *Coal, Smoke, and Sewage, Scientifically and Practically Considered*, Cave & Sever, Manchester, 1857, pp.18 and 21.

43. Voelcker, A., 'On the Injurious Effects of Smoke on Certain Building Stones and on Vegetation', *Journal of the Society of Arts*, Vol.12, 1864, p.153.

44. *Parliamentary Debates*, Vol.CCCXVIII, 1887, col.678.

45. There were around 59,000 dry-pan privies in Manchester alone in 1881. See Sharratt, A. and Farrar, K.R., 'Sanitation and Public Health in Nineteenth-century Manchester', *Memoirs of the Literary and Philosophical Society of Manchester*, Vol.114, 1971–2, p.62.

46. Mayhew, H., *op. cit.*, pp.344–5; Vendelmans, H., *The Manual of Manures*, Country Life Library, London, 1916, ch.6, 'Ashes and Soot'; and Bone, W.A., *Coal and its Scientific Uses*, Longmans, Green & Co., London, 1922, p.248.

47. Waugh, E., *Chimney Corner*, p.202.

48. *Manchester Guardian*, 28 May 1842.

49. Smith, R.A., 'On the Air of Towns', *Journal of the Chemical Society*, Vol.11, 1859, p.224.

50. Sigsworth, M. and Worboys, M., 'The Public's View of Public Health in Mid-Victorian Britain', *Urban History*, Vol.21, 1994, p.244.

51. *Ibid.*

52. See Pelling, M., 'Contagion/Germ Theory/Specificity', in Bynum, W.F. and Porter, R. (eds), *op. cit.*, ch.16.

53. Thresh, J.C., *An Enquiry into the Causes of the Excessive Mortality in No.1 District, Ancoats*, John Heywood, Manchester, 1889.

54. Sharratt, A. and Farrar, K.R., *op. cit.*, p.63.

55. Tomes, N., 'The Private Side of Public Health: Sanitary Science, Domestic Hygiene, and the Germ Theory, 1870–1900', *Bulletin of the History of Medicine*, Vol.64, 1990, p.515.

56. *Ibid.*, p.529. See also Wohl, A.S., *op. cit.*, p.89.

57. Corbin, A., *op. cit.*, p.227.

58. *Lancet*, 25 June 1892, p.1433.

59. *Manchester Guardian*, 20 June 1891.

60. Rowley, C., *Fifty Years of Work Without Wages*, Hodder & Stoughton, London, 2nd edn, n.d., pp.196–7. First published in 1912.

61. See Thresh, J.C., *op. cit.*; and Sharratt, A. and Farrar, K.R., *op. cit.*, pp.62–3.

62. Letter to the editor, in *Health Journal and Record of Sanitary Engineering*, Vol.III, 1885–6, p.37.

63. Sharratt, A. and Farrar, K.R., *op. cit.*, p.64.

64. Wohl, A.S., *op. cit.*, p.109; and Kidd, A.J., *Manchester*, Ryburn Publishing, Keele University Press, Keele, 1993, p.127.

65. Graham, J., 'Smoke Abatement in Manchester: Past and Present', *Sanitarian*, 1954, pp.236–7.

66. The phrase is Suellen Hoy's, borrowed from her excellent study *Chasing Dirt: The American Pursuit of Cleanliness*, Oxford University Press, New York, 1995. See also 'Afternoons in the Manchester Slums', in *Health Journal and Record of Sanitary Engineering*, Vol.IV, 1886–7, p.36; *Encyclopaedia Britannica*, Vol.V, Adam & Charles Black, Edinburgh, 9th edn, 1876, pp.85–6; Corlett, W.J., *The Economic Development of Detergents*, Gerald Duckworth, London, 1958; and Musson, A.E., *Enterprise in Soap and Chemicals: Joseph Crosfield and Sons, Ltd. 1815–1965*, Manchester University Press, Manchester, 1965.

67. Beeton, Mrs I., *Book of Household Management*, Ward, Lock & Co., London, 1906 edn, p.1835.

68. Horsfield, M., *Biting the Dust: The Joys of Housework*, Fourth Estate, London, 1997, p.99. For a mid-nineteenth century description of this process, see the fumigation of Mr Dolls in Dickens, C., *Our Mutual Friend*, Oxford University Press, Oxford, 1981 edn, p.538. First published in 1865.

69. *Cassell's Household Guide to Every Department of Practical Life*, Vol.IV, Cassell, Petter & Galpin, London, n.d., p.205.

70. Ross, E., *Love and Toil: Motherhood in Outcast London, 1870–1918*, Oxford University Press, New York, 1993, pp.177–8.

71. Fennings, A., *Fennings' Everbody's Doctor; or, When Ill, How to Get Well*, revised edn, n.d., pp.12–13. First published in 1864. See also *Manchester Guardian*, 1 October 1888.

72. Rolleston, C., 'The Cloud over English Life', *Westminster Review*, Vol.162, 1904, p.35. See also 'Topics of the Day: The Fog at the Throat', *Evening Standard and St. James Gazette*, 12 December 1905.

73. The phrase occurs in Sims, G.R., 'London under the Weather', in *idem* (ed.), *Living London*, Vol.II, Cassell & Co., London, 1902, p.263.

74. Taylor, W.C., *op. cit.*, p.22.

75. Brimblecombe, P., *The Big Smoke: A History of Air Pollution in London since Medieval Times*, Routledge, London, 1988, p.131.

76. Stoker, B., *Dracula*, Oxford University Press, Oxford, 1996 edn, p.116. First published in 1897.

77. Bennett, A., *Anna of the Five Towns*, Wordsworth Editions, Ware, 1994 edn, p.47. First published in 1902.

78. *Black's Guide to Manchester*, A. & C. Black, London, 1913, p.70.

79. Cook, E.T. and Wedderburn, A. (eds), *The Works of John Ruskin, Vol.I, The Poetry of Architecture*, George Allen, London, 1903, p.55. First published in book form in 1873.

80. *Ibid, Vol.XVI, 'A Joy For Ever'* and *The Two Paths*, pp.339–40. First published in 1859.

81. *Manchester Guardian*, 26 August 1887.

82. For example, see 'The Lesser Arts', in *The Collected Works of William Morris, Vol.XXII, Hopes and Fears For Art*, Routledge, London, 1992 edn. First published in 1882.

83. Morris, W., *News From Nowhere and Other Writings*, Penguin, Harmondsworth, 1993 edn, p.140. First published in 1890.

84. Engels, F., *The Condition of the Working Class in England*, Penguin, Harmondsworth, 1987 edn, p.83. First published in Germany in 1845. Also see the chapter on Manchester in *The Land We Live In: A Pictorial, Historical, and Literary Sketch-Book of the British Islands, with Descriptions of Their More Remarkable Features and Localities*, Vol.1, William S. Orr, London, n.d. (c.1847); and Briggs, A., *Victorian Cities*, Penguin, Harmondsworth, 1990 edn, ch.3.

85. Disraeli, B., *Coningsby: or the New Generation*, John Lehmann, London, 1948 edn, pp.148–9. First published in 1844.

86. *Bradshaw's Illustrated Guide to Manchester*, Bradshaw & Blacklock, Manchester, 1857, p.5.

87. For example, see John Ruskin's lecture on 'Modern Manufacture and Design' in Cook, E.T. and Wedderburn, A. (eds), *The Works of John Ruskin, Vol.XVI, The Two Paths* and *Vol. XXXIV, The Storm-Cloud of the Nineteenth Century*.

88. Clarke, A., *The Effects of the Factory System*, Grant Richards, London, 1899, p.37. George Kelsall, Littleborough, reprint of 1985.

89. *Black's Guide to Manchester*, p.20.

90. However, for some particularly ornate tall chimney designs see Rawlinson, Sir R., *Designs for Factory, Furnace, and Other Tall Chimney Shafts*, Kell Brothers, London, 1859.

91. *Ibid.*, p.1.

92. Reilly, C.H., 'The Face of Manchester', in Brindley, W.H. (ed.), *The Soul Of Manchester*, Manchester University Press, Manchester, 1929, p.100.

93. Morton, H.V., *The Call of England*, Methuen, London, 4th edn, 1929, p.144 and p.113.

94. See Douglas, M., *Purity and Danger: An Analysis of Concepts of Pollution and Taboo*, Routledge, London, 1991 edn.

95. *Manchester Guardian*, 17 August 1887.

96. Spencer, R., *A Survey of the History, Commerce and Manufactures of Lancashire*, The Biographical Publishing Company, London, 1897, p.48. See also Morton, H.V., *op. cit.*, p.174.

97. *Black's Guide to Manchester*, p.1.

98. Charles Dickens, *Hard Times*, Oxford University Press, Oxford, 1989 edn, p.166. First published in 1854.

99. For a summary, see Mosley, S., 'The "Smoke Nuisance" and Environmental Reformers in Late Victorian Manchester', *Manchester Region History Review*, Vol.X, 1996, pp.40–6.

100. *Manchester Guardian*, 28 May 1842.

101. *Ibid.*

102. *Ibid.*

103. *Ibid.*

104. Blaikie, W.G., *Better Days For Working People*, T. & A. Constable, Edinburgh University Press, Edinburgh, 1881 edn, p.130.

105. Sigsworth, E.M. (ed.), *In Search of Victorian Values: Aspects of Nineteenth-century Thought and Society*, Manchester University Press, Manchester, 1988.

106. *Parliamentary Papers*, (House of Commons), 1819 (574) VIII, p.9. For example, see *Builder*, 12 August 1899, pp.143–4.

107. *Parliamentary Papers*, (House of Commons), 1819 (574) VIII, p5 and p.18.

108. See Hamlin, C., *op. cit.*; and Wohl, A.S., *op. cit.*, p.147.

109. *Parliamentary Papers*, (House of Commons), 1845 (602) XVIII.I Part I, p.44.

110. *Manchester Guardian*, 7 June 1843. In addition, the second report of the Royal Commission on the Sanitary State of Large Towns and Populous Districts stated that 'the annual money loss to Manchester by its smoke *is double the amount of its poor rates*'. See *Parliamentary Papers*, (House of Commons), 1845 (610) XVIII.I, Part II Appendix, p.20 n.

111. *Parliamentary Papers*, (House of Commons), 1843 (583) VII, q.680.

112. *Manchester Guardian*, 28 May 1842.

113. See Chadwick, E., in Arnott, N., 'On a New Smoke-Consuming and Fuel-Saving Fire-Place', *Journal of the Society of Arts*, Vol. 2, 1854, pp.433–4.

114. *Annual Report of the Manchester and Salford Sanitary Association for 1855–6*, Powlson & Sons, Manchester, 1856, pp.6–7.

115. Russell, F.A.R., *Smoke in Relation to Fogs in London*, The National Smoke Abatement Institution, London, 1889, pp.22–6.

116. Bone, W.A., *Coal and Health*, p.6; and Manchester Air Pollution Advisory Board, *The Black Smoke Tax: An Account of Damage Done by Smoke, with an Inquiry into the Comparative Cost of Family Washing in Manchester and Harrogate*, Henry Blacklock & Co., Manchester, 1920, pp.5–12.

117. *Ibid.*, pp.1–2.

118. *Times*, 2 January 1855.

119. Spence, P., *op. cit.*, pp.19–20 and p.12.

120. Rawnsley, H.D., 'Sunlight or Smoke?' *Contemporary Review*, Vol.57, 1890, p.521.

121. Coal Smoke Abatement Society, *The Coal Smoke Nuisance: Why It Should Be Abated*, Pamphlet, London, 1913, p.3.

122. Lodge, O., 'Electricity and Gas v. Smoke', *Exhibition Review*, No.6, 1882, p.2. For earlier criticisms of the open coal fire see, Arnott, N., *op. cit.*

123. *Builder*, 12 August 1899, p.144.

124. Smoke Abatement League of Great Britain, *Proposed Branch for Manchester and District*, Pamphlet, c.1910, p.4.

125. *Builder*, 12 August 1899, p.143.

126. Pepper, D., *Modern Environmentalism: An Introduction*, Routledge, London, 1996, pp.230–3; and Worster, D., *Nature's Economy: A History of Ecological Ideas*, Cambridge University Press, New York, 2nd edn, 1994, p.303.

127. Arnott, N., *op. cit.*, p.428.

128. Percy, J., 'Coal and Smoke', *Quarterly Review*, Vol.119, 1866, p.463 and p.456.

129. *Ibid.*, p.450. See also Smith, C. and Wise, M.N., *Energy and Empire: A Biographical Study of Lord Kelvin*, Cambridge University Press, Cambridge, 1989.

130. Clapp, B.W., *An Environmental History of Britain since the Industrial Revolution*, Longman, London, 1994, pp.152–6.

131. Jevons, W.S., *The Coal Question*, 3rd edn, 1906, p.459. Augustus M. Kelley, New York, reprint of 1965.

132. See Bevan, P., 'Our National Coal Cellar', *Gentleman's Magazine*, Vol.9, 1872, pp.277–8.

133. *Ibid.*, p.278.

134. *Manchester Guardian*, 9 November 1889.

135. Bevan, P., *op. cit.*, p.277.

136. 'The Smoke of our Burning', *Spectator*, 27 November 1926, p.950.

137. Evelyn, J., *Fumifugium*, W. Godbid, London, 1661. Rota, Exeter, reprint of 1976, prefatory notes.

138. *Parliamentary Papers*, (House of Commons), 1843 (583) VII, qs. 680 and 686.

139. *Parliamentary Papers*, (House of Commons), 1845 (610) XVIII.I, Part II Appendix, pp.113–16.

140. *Ibid.*, p.112.

141. *Ibid.*

142. Cullen, M.J., *The Statistical Movement in Early Victorian Britain: The Foundations of Empirical Social Research*, The Harvester Press, Hassocks, 1975, p.146.

143. Statistical inquiries were commonly believed to tell a 'truthful', objective story about the state of the nation's lower classes, even if their main conclusions are today recognised to be largely 'anticipated and preconceived'. See *Ibid.*; Wohl, A.S., *op.*

cit., p.145; and Fraser, D., *The Evolution of the British Welfare State: A History of Social Policy since the Industrial Revolution*, Macmillan, London, 1992 edn.

144. Leigh, J., 'Coal Smoke: Report to the Health and Nuisance Committees of the Corporation of Manchester', *Health Journal and Record of Sanitary Engineering*, Vol.1, 1883, p.52.

145. *Ibid.*

146. What is more, the contrast between town and country would have appeared even plainer if the death rates for South Lancashire had been disaggregated from those of Cheshire.

147. Leigh, J., 'Coal Smoke', p.69.

148. Ransome, A., 'The Smoke Nuisance. (A Sanitarian's View)', *Exhibition Review*, No.1, 1882, p.3.

149. *Annual Report of the Manchester and Salford Sanitary Association for 1876*, Powlson & Sons, Manchester, 1877, p.9.

150. Coal Smoke Abatement Society, *The Coal Smoke Nuisance*, p.1.

151. See Tables 1, 2, and 3 in Hardy, A., 'Rickets and the Rest: Child-care, Diet and the Infectious Children's Diseases, 1850–1914', *Social History of Medicine*, Vol.5, 1992, pp.390–1.

152. *Parliamentary Papers*, (House of Commons), 1845 (610) XVIII.I, Part II Appendix, p.112.

153. Leigh, J., 'Coal Smoke', p.52.

154. *Ibid.* See also Thompson, F.M.L., *The Rise of Respectable Society: A Social History of Victorian Britain, 1830–1900*, Fontana Press, London, 1988, pp.117–18.

155. Fyfe, P., 'The Pollution of the Air', p.75.

156. M'Lean, H.A., 'The Smokeless City: A Retrospect and a Prospect', in Smoke Abatement League of Great Britain, *Lectures Delivered in the Technical College, Glasgow 1910–1911*, Corporation of Glasgow, Glasgow, 1912, p.59.

157. Fyfe, P., 'The Pollution of the Air', p.75.

158. *Ibid.*

159. *Ibid.*, p.71 and pp.73–4.

160. Brend, W.A., *Health and the State*, Constable & Co., London, 1917, p.87.

161. Littlejohn, H., *Report on the Causes and Prevention of Smoke from Manufacturing Chimneys*, Wm. Townsend & Son, Sheffield, 1897, p.22. See also Niven, J. and Tattersall, C.H., 'Meteorology and Health of Manchester and Salford', in Ray, J.H. (ed.), *British Medical Association. Manchester Meeting 1902. Handbook and Guide to Manchester*, F. Ireland, Manchester, 1902, pp.44–6.

162. 'Afternoons in the Manchester Slums', in *Health Journal and Record of Sanitary Engineering*, Vol.IV, 1886–7, p.56; Evans, R.J., *Death in Hamburg: Society and Politics in the Cholera Years 1830–1910*, Penguin, Harmondsworth, 1990 edn, pp.226–37; and Oddy, D.J., 'The Health of the People', in Barker, T. and Drake, M. (eds), *Population and Society in Britain 1850–1980*, Batsford Academic & Educational, London, 1982, pp.122–3.

163. Wohl, A.S., *op. cit.*, pp.118–19; and Morris, R.J., *Cholera 1832: The Social Response to an Epidemic*, Croom Helm, London, 1976.

164. Evans, R.J., *op. cit.*, pp.228–30.

165. Carpenter, E., 'The Smoke-Plague and its Remedy', *Macmillan's Magazine*, Vol. 62, 1890, pp.204–6.

166. *Manchester Guardian*, 25 June 1888.

167. See discussion in Smith, R.A., 'What Amendments', p.537.

168. *Lucretia*, 'Smoke, and its Effects on Health', *Exhibition Review*, No.5, 1882, p.2. See also Ernest Hart's evidence to the Select Committee investigating Smoke Nuisance Abatement in 1887, *Parliamentary Papers*, (House of Lords), 1887 (321) XII, p.40; and Chubb, L.W., *Smoke Abatement*, Coal Smoke Abatement Society, Pamphlet, London, c.1912, p.13.

169. *Manchester Guardian*, 25 April 1902.

170. Porter, D., '"Enemies of the Race": Biologism, Environmentalism, and Public Health in Edwardian England', *Victorian Studies*, Vol.34, 1991, pp.159–78.

171. Ransome, A., 'The Smoke Nuisance', p.3. See also Ellison, R.C., 'On the Influence of the Purity or Impurity of the External Air on the Health and Moral Tendencies of a Dense Population', *Transactions of the Sanitary Institute of Great Britain*, Vol.IV, 1882–3, p.224.

172. Lucretia, 'Smoke', pp. 2–3.

173. Chinn, C., *Poverty amidst Prosperity: The Urban Poor in England, 1834–1914*, Manchester University Press, Manchester, 1995, p.114.

174. *Ibid.*

175. *Manchester Guardian* 25 April 1902.

176. Scott, F., 'The Case for a Ministry of Health', *Transactions of the Manchester Statistical Society*, 1902–3, p.99.

177. Porter, B., *The Lion's Share: A Short History of British Imperialism 1850–1983*, Longman, Harlow, 2nd edn, 1984, pp.116–18.

178. See Table 12 in Kennedy, P., *The Rise and Fall of the Great Powers: Economic Change and Military Conflict from 1500 to 2000*, Fontana Press, London, 1988, p.255.

179. Porter, B., *op. cit.*, p.130.

180. Barr, Sir J., 'The Advantages, from a National Standpoint, of Compulsory Physical Training of the Youth of this Country', in *Manchester and Salford Sanitary Association Annual Report 1914*, Sherratt & Hughes, Manchester, 1915, p.22.

181. Scott, F., 'The Case for a Ministry of Health', p.100 and p.141; and Froude, J.A., *Oceana, or England and her Colonies*, Longmans, Green & Co., London, 1886, pp.7–10 and pp.331–4.

182. Emrys-Jones, A., 'Smoke and Impure Air', *Exhibition Review*, No.4, 1882, p.3.

183. Horsfall, T.C., 'The Social Aspect of the Smoke Question', *Exhibition Review*, No.1, 1882, p.2; and Wilfred Lely in *Manchester Guardian* 22 August 1887.

184. Ellison, R.C., *op. cit.*, p.225.

185. *Ibid.*, p.224.

186. Horsfall, T.C., 'The Government of Manchester', p.10.

187. *Ibid.*, pp.18–25.

188. Quoted in *Parliamentary Papers*, (House of Commons), 1904 (Cd.2175) XXXII.I, p.20.

189. Horsfall, T.C., 'The Social Aspect', p.2; and *idem*, 'The Government of Manchester', pp.4–6.

190. G., J.M., 'The Smoke Nuisance: A Lady's View', *Exhibition Review*, No.3, 1882, p.2; and Kidd, A.J., *Manchester*, pp.147–8.

191. *Parliamentary Papers*, (House of Commons), 1845 (610) XVIII.I, Part II Appendix, p.112.

192. Quoted in Scott, F., 'The Need for Better Organisation of Benevolent Effort in Manchester and Salford', *Transactions of the Manchester Statistical Society*, 1884–85, p.141 and p.143.

193. *Parliamentary Papers*, (House of Commons), 1904 (Cd.2175) XXXII.I, p.20.

194. *Ibid.*

195. *Ibid.*, p.86.

196. Horsfall, T.C., *The Place of 'Admiration, Hope, and Love' in Town Life: Reply by T.C. Horsfall to an Address presented to him on 30th June, 1910, by nearly three hundred of his Fellow-Citizens*, Pamphlet, Manchester, 1910, pp.9–24. For useful discussions of two such charitable initiatives see Harrison, M., 'Art and Philanthropy: T.C. Horsfall and The Manchester Art Museum', in Kidd, A.J. and Roberts, K.W. (eds), *City, Class and Culture: Studies of Social Policy and Cultural Production in Victorian Manchester*, Manchester University Press, Manchester, 1985; and Kay, A., 'Charles Rowley and the Ancoats Recreation Movement, 1876–1914', *Manchester Region History Review*, Vol.VII, 1993, pp.45–54.

197. Disraeli, B., *op. cit.*, pp.148–9.

198. Faucher, L., *Manchester in 1844; Its Present Condition and Future Prospects*, Abel Heywood, Manchester, 1844, pp.16–25. Frank Cass, London, reprint of 1969.

199. Kidd, A.J., 'The Industrial City and its Pre-industrial Past: the Manchester Royal Jubilee Exhibition of 1887', *Transactions of the Lancashire and Cheshire Antiquarian Society*, Vol.89, 1993, p.58.

200. Seed, J., '"Commerce and the Liberal Arts": The Political Economy of Art in Manchester, 1775–1860', in Wolff, J. and Seed, J. (eds), *The Culture of Capital: Art, Power and the Nineteenth-century Middle Class*, Manchester University Press, Manchester, 1988, p.67.

201. Horsfall, T.C., 'The Government of Manchester', p.12.

202 Kidd, A.J., *Manchester*, pp.72–9 and pp.159–64.

203. Cook, E.T. and Wedderburn, A. (eds), *The Works of John Ruskin, Vol.XXIX, Fors Clavigera*, p.224.

204. Brooks, M.W., *John Ruskin and Victorian Architecture*, Thames & Hudson, London, 1989, p.173.

205. *Manchester Guardian*, 22 and 27 August 1887.

206. Horsfall, T.C., 'The Government of Manchester', pp.6–7. See also Kidd, A.J., 'The Industrial City', pp.72–3.

207. Gould, P.C., *Early Green Politics: Back to Nature, Back to the Land, and Socialism in Britain, 1880–1900*, Harvester Press, Brighton, 1988, p.39.

208. Harrison, J.F.C., *Late Victorian Britain, 1875–1901*, Fontana Press, London, 1990, p.148.

209. *Ibid.*

210. Blatchford, R., *Merrie England*, Clarion Newspaper Company, London, 1895 edn, pp.21–3.

211. Copy in the collection of Stephen Mosley. See also Gould, P.C., *op. cit.*, p.41.

212. See the list of suggested reading in Robert Blatchford's *Merrie England*, pp.205–6.

213. Clarke, A., *The Effects of the Factory System*, p.173.

214. *Ibid.*, p.174.

215. Gould, P.C., *op. cit.*, pp.36–7.

216. Clarke, A., *The Effects of the Factory System*, p.38.

217. Bone, W.A., *Coal and Health*, p.15.

218. Mayhew, H., *op. cit.*, pp.81–2.

219. Roberts, R., *op. cit.*, p.71.

220. I have adapted this concept from Potter, J. and Wetherell, M., *Discourse and Social Psychology: Beyond Attitudes and Behaviour*, Sage Publications, London, 1994 edn, chapter 2.

221. *Parliamentary Papers*, (House of Commons), 1843 (583) VII, qs.1553 and 1554. See also the evidence of Thomas Cockshott Rusher, Nuisance Inspector at Leeds, in *Parliamentary Papers*, (House of Commons), 1845 (289) XIII, q.373; and the observations of the Reverend Mr Clay, Prison Chaplain at Preston, in Arnott, N., *op. cit.*, p.434.

222. There is substantial evidence of working-class interest in improving environmental conditions in late-Victorian Manchester and Salford. For example, Michael Harrison has pointed to the formation of several Workmen's Sanitary Associations and Healthy Homes Societies in the twin cities that attracted genuine working-class support. See *idem*, 'Housing and Town Planning in Manchester before 1914', in Sutcliffe, A. (ed.), *British Town Planning: The Formative Years*, Leicester University Press, Leicester, 1981, p.112.

223. *Manchester Guardian*, 12 December 1882.

224. *Annual Report of the Manchester and Salford Noxious Vapours Abatement Association for 1883* in *Annual Report of the Manchester and Salford Sanitary Association for 1883*, p.82. Manchester Central Library Ref., 614.06 M1.

225. Walton, J.K., *op. cit.*, chapter 10.

226. Rawnsley, H.D., *op. cit.*, p.523. Moreover, Edward Carpenter recorded the hostile comments of Sheffield's workers as he campaigned against smoke in the 1890s. See, *Daily News and Leader,* 28 May 1921.

227. Clarke, A., *Moorlands and Memories,* p.9.

Part Three: The Search for Solutions

1. Clarke, A., *The Effects of the Factory System,* Grant Richards, London, 1899, p.22. George Kelsall, Littleborough, reprint of 1985.

2. Rein, M. and Schön, D.A., 'Frame-Reflective Policy Discourse', *Beleidsanalyse,* Vol.15, No.4, 1986, pp.4–18; and Hajer, M.A., *The Politics of Environmental Discourse: Ecological Modernisation and the Policy Process,* Clarendon Press, Oxford, 1995, p.62.

3. *Ibid.*

4. Dickens, C., 'Smoke or No Smoke', *Household Words,* Vol.9, 1854, pp.464–5.

5. For example, see Rawnsley, H.D., 'Sunlight or Smoke?' *Contemporary Review,* Vol.57, 1890, pp.512–24.

6. Dickens, C., *op. cit.,* p.465.

7. Chadwick, E., *Report on the Sanitary Condition of the Labouring Population of Great Britain,* 1842, edited by M.W. Flinn, Edinburgh University Press, Edinburgh, 1965, pp.355–8; *Manchester Guardian,* 28 May 1842; and Kargon, R.H., *Science in Victorian Manchester: Enterprise and Expertise,* Manchester University Press, Manchester, 1977, pp.112–16. Smoke abatement often appears to have enjoyed a higher profile when a city was about to host a prestigious meeting or exhibition. For example, see Rosen, C.M., 'Businessmen Against Pollution in Late Nineteenth Century Chicago', *Business History Review,* Vol. 69, 1995, pp.351–97.

8. *Manchester Guardian,* 28 May 1842.

9. *Parliamentary Papers,* (House of Commons), 1843 (583) VII, q.680.

10. *Manchester Guardian,* 28 May 1842.

11. For a recent discussion of this topic see Williams, R. and Edge, D., 'The Social Shaping of Technology', *Research Policy,* Vol. 25, 1996, pp.865–99.

12. *Ibid.,* p.866.

13. Bijker, W.E. and Law, J. (eds), *Shaping Technology/Building Society: Studies in Sociotechnical Change,* MIT Press, Cambridge, Massachusetts, 1992, p.3.

14. Cardwell, D., *The Fontana History of Technology,* Fontana Press, London, 1994; Musson, A.E., 'Industrial Motive Power in the United Kingdom, 1800–70', *Economic History Review,* 2nd Series, Vol.XXIX, 1976, pp.415–39; and Landes, D.S., *The Unbound Prometheus: Technological Change and Industrial Development in Western Europe from 1750 to the Present,* Cambridge University Press, New York, 1969.

15. Walton, J.K., *Lancashire: A Social History, 1558–1939,* Manchester University Press, Manchester, 1990 edn, p.103.

16. Williams, C.W., *The Combustion of Coal and the Prevention of Smoke Chemically and Practically Considered*, John Weale, London, 1854, p.ii.

17. This section owes much to Carlos Flick's work on smoke abatement technology. See *idem*, 'The Movement for Smoke Abatement in 19th-Century Britain', *Technology and Culture*, Vol. 21, 1980, pp.29–50.

18. *Parliamentary Papers*, (House of Commons), 1843 (583) VII, q.1001; and Ashby, E. and Anderson, M., 'Studies in the Politics of Environmental Protection: The Historical Roots of the British Clean Air Act, 1956: I. The Awakening of Public Opinion over Industrial Smoke, 1843–53', *Interdisciplinary Science Reviews*, Vol. 1, 1976, p.283.

19. Flick, C., *op. cit.*, pp.39–46.

20. *Ibid.*, p.45.

21. For example, see the views of the industrialist Sir Josiah Guest MP in *Parliamentary Papers*, (House of Commons), 1845 (489) XIII, q.734.

22. *Parliamentary Papers*, (House of Commons), 1820 (244) II, p.15.

23. *Parliamentary Papers*, (House of Commons), 1819 (574) VIII, Appendix No. 4, p.18.

24. *Manchester Guardian*, 11 June 1842.

25. See Winstanley, M., 'The Factory Workforce', in Rose, M.B. (ed.), *The Lancashire Cotton Industry: A History Since 1700*, Lancashire County Books, Preston, 1996.

26. Richardson, C.J., *The Smoke Nuisance and Its Remedy. With Remarks on Liquid Fuel*, Atchley & Co., London, 1869, pp.45–6; and Flick, C., *op. cit.*, pp.44–8.

27. For example, John Juckes claimed that by using his 'moving grates' industrialists could save between 75–80 per cent on their fuel bills. See *Parliamentary Papers*, (House of Commons), 1843 (583) VII, q.751.

28. *Parliamentary Papers*, (House of Commons), 1845 (489) XIII, q.937.

29. Gatrell, V.A.C., 'Labour, Power, and the Size of Firms in Lancashire Cotton in the Second Quarter of the Nineteenth Century', *Economic History Review*, 2nd Series, Vol. XXX, 1977, pp.95–129.

30. See *Parliamentary Papers*, (House of Commons), 1854 LXI 533.

31. Gatrell, V.A.C., *op. cit.*, p.120.

32. *Ibid.*

33. Beilby, G., 'President's Address', *Journal of the Society of Chemical Industry*, Vol. XVIII, 1899, p.644.

34. Clapp, B.W., *An Environmental History of Britain since the Industrial Revolution*, Longman, Harlow, 1994, p.21.

35. Ashby, E. and Anderson, M., *op. cit.*, p.283.

36. See 'Instructions to Firemen' issued on 31 March 1856 in Manchester Steam Users' Association, *A Sketch of the Foundation and of the Past Fifty Year's Activity of the Manchester Steam Users' Association for the Prevention of Steam Boiler Explosions and*

for the Attainment of Economy in the Application of Steam, Taylor, Garnett, Evans & Co., Manchester, 1905, p.30; and *Manchester Guardian*, 11 June 1842.

37. Mayhew, H., *London Labour and the London Poor; A Cyclopædia of the Condition and Earnings of Those that Will Work, Those that Cannot Work, and Those that Will not Work*, Vol.2, Dover Publications, New York, 1968, p.87.

38. For example, see *Parliamentary Papers*, (House of Commons), 1845 (489) XIII, q.943.

39. *Manchester Guardian*, 28 May 1842. See also: *Parliamentary Papers*, (House of Commons), 1843 (583) VII, qs.680 and 700.

40. *Parliamentary Papers*, (House of Commons), 1845 (289) XIII, q.385.

41. *Parliamentary Papers*, (House of Commons), 1843 (583) VII, q.697.

42. *Ibid.*, q. 1044; and Flick, C., *op. cit.*, p.49.

43. *Parliamentary Papers*, (House of Commons), 1843 (583) VII, q.861; and Manchester and District Regional Smoke Abatement Committee, *Smoke Abatement and Fuel Economy in Steam Boiler Practice*, Pamphlet, Manchester, n.d., c.1921, pp.4–6.

44. *Parliamentary Papers*, (House of Commons), 1843 (583) VII, qs.54–9 and 350–2; Manchester Steam Users' Association, *op. cit.*; and Musson, A.E., *op. cit.*, p.421.

45. *Manchester Guardian*, 11 June 1842; *Parliamentary Papers*, (House of Commons), 1843 (583) VII, q.1001.

46. Manchester Steam Users' Association, *op. cit.*, p.7 and p.16.

47. Fowler, A. and Wyke, T. (eds), *The Barefoot Aristocrats: A History of the Amalgamated Association of Cotton Spinners*, George Kelsall, Littleborough, 1987, Appendix V, 'Wages of Spinners, Other Textile Workers, Crafts, Labourers Etc., in Cotton Towns: Selected Years 1810–59', p.245; and Ashby, E. and Anderson, M., *op. cit.*, p.282.

48. Generally speaking, industrialists on the continent also made more efficient use of their coal resources due to its higher cost. See Landes, D.S., *op. cit.*, pp.180–1.

49. *Manchester Guardian*, 11 June 1842; *Parliamentary Papers*, (House of Commons), 1843 (583) VII, qs.1013–22; Manchester Steam Users' Association, *op. cit.*, p.14; and Cardwell, D., *op. cit.*, pp.213–14.

50. Alloway, B.J. and Ayres, D.C., *Chemical Principles of Environmental Pollution*, Blackie Academic & Professional, Glasgow, 1993, table 6.1, p.200. See also, *Parliamentary Papers*, (House of Commons), 1843 (583) VII, q.1015; National Smoke Abatement Society, *The National Smoke Abatement Handbook*, Service Guild, Manchester, 1931, p.25; and Brimblecombe, P., *The Big Smoke: A History of Air Pollution in London since Medieval Times*, Routledge, London, 1988, p.98.

51. *Manchester Guardian*, 11 June 1842; and Tunzelmann, N. von, 'Coal and Steam Power', in Langton, J. and Morris, R.J. (eds), *Atlas of Industrialising Britain, 1780–1914*, Methuen, London, 1986, pp.72–4.

52. *Ibid.*, maps 8.1–8.4.

53. Lowe, J., 'In Praise of Tall Chimneys', *Industrial Heritage*, Vol.8, 1989, p.14.

54. *Ibid.*, p.15.

55. *Manchester Guardian,* 11 June 1842. However, in 1844 the Manchester engineer William Fairbairn, who attended MAPS' meetings, invented the Lancashire boiler. It was designed with the furnace encased inside the boiler to help improve the transfer of heat from the fire, and it was still in widespread use in the mid-twentieth century. See Clapp, B.W., *op. cit.,* p.164.

56. *Parliamentary Papers,* (House of Commons), 1843 (583) VII, qs.369–70; and *Parliamentary Papers,* (House of Commons), 1844 (572) XVII.I, qs.3918–26.

57. *Parliamentary Papers,* (House of Commons), 1843 (583) VII, q.1524.

58. For example, see the evidence of the noted sanitary reformer Dr Neil Arnott to the Royal Commission on the Sanitary State of Large Towns and Populous Districts in *Parliamentary Papers,* (House of Commons), 1844 (572) XVII.I, qs.3893–3981.

59. Ransome, A., 'On Foul Air and Lung Disease', in *Health Lectures for the People,* Vol.1, Manchester and Salford Sanitary Association, Manchester, 1878, p.50. See also Brewer, Rev. Dr E.C., *A Guide to the Scientific Knowledge of Things Familiar,* Jarrold & Sons, London, 46th edn, 1897, chapter 19. First published in 1848, Dr Brewer's book had sold 300,000 copies by 1897.

60. See Wager, H. and Herbert, A., 'Bad Air and Bad Health', *Contemporary Review,* Vol.59, 1891, pp.852–74.

61. Godwin, G., *Town Swamps and Social Bridges. The Sequel of 'A Glance at the Homes of the Thousands',* Routledge, Warnes, & Routledge, London, 1859, p.102. Leicester University Press, New York, reprint of 1972.

62. Brewer, Rev. Dr E.C., *op. cit.,* p.267.

63. Fox, W., *The Working Man's Model Family Botanic Guide; or, Every Man His Own Doctor,* William Fox & Sons, Sheffield, 12th edn, 1889, p.83. First published in 1852, Dr Fox's book had sold some 50,000 copies by 1889.

64. *Ibid.*, pp.82–3.

65. Ransome, A., 'Foul Air', p.48.

66. *Parliamentary Papers,* (House of Commons), 1844 (572) XVII.I, q.275.

67. 'The Smoke Abatement Exhibition', *Nature,* Vol. 25, 1882, p.219.

68. Ransome, A., *op. cit.,* p.58. See also Thresh, J.C., *An Enquiry into the Causes of the Excessive Mortality in No.1 District, Ancoats,* John Heywood, Manchester, 1889, pp.28–30.

69. Arnott, N., 'On a New Smoke-Consuming and Fuel-Saving Fire-Place', *Journal of the Society of Arts,* Vol. 2, 1854, pp.428–34.

70. *Parliamentary Papers,* (House of Commons), 1843 (583) VII, q.2062.

71. For example, see Dixon Mann, J., 'Cottage Ventilation and its Influence on Health', *Health Lectures for the People,* New Series No.2, Manchester and Salford Sanitary Association, Sherratt & Hughes, Manchester, 1901.

72. Turner, M.J., 'Before the Manchester School: Economic Theory in Early Nineteenth-century Manchester', *History,* Vol. 79, 1994, pp.216–41; Gatrell, V.A.C.,

'Incorporation and the Pursuit of Liberal Hegemony in Manchester 1790–1839', in Fraser, D. (ed.), *Municipal Reform in the Industrial City*, Leicester University Press, Leicester, 1982; and Kidd, A.J., *Manchester*, Ryburn Publishing, Keele University Press, Keele, 1993, pp.71–2.

73. *Manchester Guardian*, 28 May 1842.

74. For example, see Henry Houldsworth's letter of August 1845 to the Mayor of Manchester. Borough of Manchester, *Proceedings of the Council*, 1844–45, p.193. Manchester Central Library Ref., 352.042 M22.

75. *Parliamentary Papers*, (House of Commons), 1843 (583) VII, qs.707 and 1050.

76. *Manchester Guardian*, 28 May 1842.

77. *Parliamentary Papers*, (House of Commons), 1843 (583) VII, q.702.

78. *Ibid.*, qs.702–3.

79. *Manchester Guardian*, 28 May 1842.

80. *Parliamentary Papers*, (House of Commons), 1843 (583) VII, q.697.

81. *Ibid.*, q.701.

82. *Manchester Guardian*, 7 June 1843.

83. See, Rosen, C., 'Differing Perceptions of the Value of Pollution Abatement across Time and Place: Balancing Doctrine in Pollution Nuisance Law, 1840–1906', *Law and History Review*, Vol.11, 1993, pp.303–81; McLaren, J.P.S., 'Nuisance Law and the Industrial Revolution – Some Lessons from Social History', *Oxford Journal of Legal Studies*, Vol.3, 1983, pp.155–221; and Brenner, J.F., 'Nuisance Law and the Industrial Revolution', *Journal of Legal Studies*, Vol.III, 1974, pp.403–33.

84. McLaren, J.P.S., *op. cit.*, p.169.

85. Rosen, C., *op. cit.*, p.303.

86. Brenner, J.F., *op. cit.*, p.414.

87. *Ibid.* See also Brackenbury, C.E., *Some Legal Aspects of the Smoke Nuisance*, pamphlet, King, Sell, & Olding, London, n.d., c.1910, p.7.

88. Rosen, C., *op. cit.*, p.303.

89. McLaren, J.P.S., *op. cit.*, p.170 n.

90. Rosen, C., *op. cit.*, p.320.

91. Brenner, J.F., *op. cit.*, p.432.

92. McLaren, J.P.S., *op. cit.*, p.160.

93. *Ibid.*, p.196.

94. Smith, R.A., 'Some Remarks on the Air and Water of Towns', *Philosophical Magazine*, Vol. 30, 1847, pp.478–82.

95. See the trial reports of *Regina vs. Spence* in *Manchester Guardian*, 24 and 25 August 1857. Also see Roderick, G.W. and Stephens, M.D., 'Profits and Pollution: Some Problems facing the Chemical Industry in the Nineteenth Century. The Corporation of Liverpool versus James Muspratt, Alkali Manufacturer, 1838', *Industrial Archaeology*, Vol. 11, 1974, pp.35–45.

96. *Manchester Guardian*, 24 August 1857.

97. *Ibid.*

98. Roderick, G.W. and Stephens, M.D., *op. cit.*, pp.41–2; and Percy, J., 'Coal and Smoke', *Quarterly Review*, Vol.119, 1866, p.440. Indeed, during the Spence case the noted Victorian scientist Edward Frankland caused much controversy. Initially, he acted as a paid consultant for the defendant, only to change his allegiance later to give evidence for the prosecution. See *Manchester Guardian*, 24 and 25 August 1857; and Kargon, R.H., *op. cit.*, pp.140–1.

99. McLaren, J.P.S., *op. cit.*, p.197.

100. *Ibid.*, pp.195–7; Roderick, G.W. and Stephens, M.D., *op. cit.*, pp.42–3; Dingle, A.E., '"The Monster Nuisance of All": Landowners, Alkali Manufacturers, and Air Pollution, 1828–64', *Economic History Review*, 2nd Series, Vol. XXXV, 1982, pp.529–48; and Rees, R., *King Copper: South Wales and the Copper Trade 1584–1895*, University of Wales Press, Cardiff, 2000.

101. McLaren, J.P.S., *op. cit.*, pp.215–16. However, some landed families did respond positively to the industrial opportunities of the era, being 'aggressively competitive entrepreneurs in their own right'. See Walton, J.K., *op. cit.*, p.127.

102. Wheeler, J., *Manchester: Its Political, Social and Commercial History, Ancient and Modern*, Whittaker & Co., London, 1836, pp.268–9; and Baker, H., 'On the Growth of the Commercial Centre of Manchester, Movement of Population, and Pressure of Habitation, – Census decenniad 1861–71', *Transactions of the Manchester Statistical Society*, 1871–2, pp.93–4.

103. Kidd, A.J., *op. cit.*, p.65; and Gatrell, V.A.C., 'Incorporation', p.22.

104. McLaren, J.P.S., *op. cit.*, pp.200–1; Webb, S. and Webb, B., *English Local Government from the Revolution to the Municipal Corporations Act: The Manor and the Borough*, part 1, Longmans, Green & Co., London, 1908, pp.21–30 and pp.99–113; and Redford, A. and Russell, I.S., *The History of Local Government in Manchester. Volume I: Manor and Township*, Longmans, Green & Co., London, 1939, part 1, 'Manchester Under Manorial Rule'.

105. Webb, S. and Webb, B., *op. cit.*, pp.23–4.

106. Earwaker, J.P., *The Court Leet Records of Manchester*, 1552–1846, Volume IX, Henry Blacklock, Manchester, 1889; and Vigier, F., *Change and Apathy: Liverpool and Manchester during the Industrial Revolution*, MIT Press, Cambridge, Massachusetts, 1970, pp.128–9.

107. Earwaker, J.P., *op. cit.*, Vols.IX–XII; and Vigier, F., *op. cit.*, p.129 n.

108. McLaren, J.P.S., *op. cit.*, p.217.

109. Gatrell, V.A.C., 'Incorporation', p.30.

110. Kidd, A.J., *op. cit.*, p.65.

111. Lloyd-Jones, R. and Lewis, M.J., *Manchester and the Age of the Factory: The Business Structure of Cottonopolis in the Industrial Revolution*, Croom Helm, Beckenham, 1988, pp.135–7.

112. Webb, S. and Webb, B., *op. cit.*, pp.110–11; and Redford, A. and Russell, I.S., *op. cit.*, pp.232–3 and p.258.

113. Redford, A. and Russell, I.S., *The History of Local Government in Manchester. Volume II: Borough and City*, Longmans, Green & Co., London, 1940, p.75.

114. Redford, A. and Russell, I.S., *Manor*, p.219.

115. *Ibid.*, pp.219–20.

116. *Parliamentary Papers*, (House of Commons), 1843 (583) VII, q.1544.

117. *Parliamentary Papers*, (House of Commons), 1846 (194) XLIII, Appendices A and B.

118. 7 & 8 Vict., Cap. xl., Clause LXXV.

119. *Parliamentary Papers*, (House of Commons), 1843 (583) VII, qs.728–33.

120. *Parliamentary Debates*, Vol.CVII, 1849, col.204.

121. *Parliamentary Papers*, (House of Commons), 1846 (194) XLIII, p.5.

122. *Ibid.*, pp.4–5.

123. *Parliamentary Papers*, (House of Commons), 1843 (583) VII, Appendix No.6, 'Police Report on Smoke Emitted from the Factory Chimneys in Manchester'.

124. Brenner, J.F., *op. cit.*, p.428.

125. See *Draft Minutes of the Manchester Corporation Nuisance Committee*, 23 February 1859 and 7 February 1883. Manchester Central Library Refs., M.9/65/1/2, and M9/65/1/10. Also see Graham, J.W., *The Destruction of Daylight: A Study in the Smoke Problem*, George Allen, London, 1907, p.119.

126. Manchester and Salford Noxious Vapours Abatement Association, *Minute Book, 1876–1891*, (hereafter NVAA, *Minutes*), committee meeting, 23 September 1881. Manchester Central Library Ref., M.126/6/1/1.

127. *Parliamentary Debates*, Vol. LXXVI, 1844, col.284; and *Parliamentary Papers*, (House of Commons), 1846 (194) XLIII, p.5.

128. Flick, C., *op. cit.*, p.36.

129. *Parliamentary Papers*, (House of Commons), 1846 (194) XLIII, p.5.

130. *Ibid.*, p.4.

131. *Ibid.*

132. Walton, J.K., *op. cit.*, p.134; and Kidd, A.J., 'Introduction: The Middle Class in Nineteenth-century Manchester', in Kidd, A.J. and Roberts, K.W. (eds), *City, Class and Culture: Studies of Social Policy and Cultural Production in Victorian Manchester*, Manchester University Press, Manchester, 1985.

133. Gatrell, V.A.C., 'Incorporation', p.31.

134. *Parliamentary Papers*, (House of Commons), 1843 (583) VII, Appendix No.6, 'Police Report on Smoke Emitted from the Factory Chimneys in Manchester'.

135. Walton, J.K., *op. cit.*, p.134.

136. *Parliamentary Papers*, (House of Commons), 1843 (583) VII, q.745.

137. Ashby, E. and Anderson, M., *op. cit.*, pp.283–7.

138. *Ibid.*, p.287.

139. Kidd, A.J., *Manchester*, p.45.

140. *Parliamentary Papers*, (House of Commons), 1843 (583) VII, q.55. It is worth noting that Muntz was one of Birmingham's major smoke polluters. See McLaren, J.P.S., *op. cit.*, p.218.

141. Ure, A., *The Philosophy of Manufactures: or, An Exposition of the Scientific, Moral, and Commercial Economy of the Factory System of Great Britain*, H.G. Bohn, London, 3rd edn, 1861, p.31. First published in 1835. Jeremy, D.J., 'Lancashire and the International Diffusion of Technology', in Rose, M.B. (ed.), *The Lancashire Cotton Industry: A History Since 1700*, Lancashire County Books, Preston, 1996, pp.210–12.

142. See Treblicock, C., *The Industrialisation of the Continental Powers, 1780–1914*, Longman, Harlow, 1989 edn.

143. *Parliamentary Debates*, Vol. LXXVIII, 1845, col.1368; and *Parliamentary Debates*, Vol. CVII, 1849, cols.193–207.

144. *Parliamentary Debates*, Vol. CV, 1849, col. 1261.

145. *Ibid.*, col. 1262.

146. *Parliamentary Debates*, Vol. CVII, 1849, col. 197.

147. Belchem, J., *Industrialisation and the Working Class: The English Experience, 1750–1900*, Scolar Press, Aldershot, 1991, pp.33–6; and Hobsbawm, E.J., *Labouring Men: Studies in the History of Labour*, Weidenfield & Nicolson, London, 1964, Table I, p.74.

148. Waugh, E., *Lancashire Sketches*, Second Series, John Heywood, Manchester, 1892, p.182. Although Waugh was speaking about conditions during the Cotton Famine of the 1860s, rather than the economic distress of the 1840s, he expressed a view long held by many of Lancashire's working classes.

149. For example, see *Parliamentary Papers*, (House of Commons), 1843 (583) VII, q.271.

150. *Ibid.*, qs.865 and 1124.

151. *Parliamentary Papers*, (House of Commons), 1846 (194) XLIII, p.3.

152. *Ibid.*, Appendix, pp.10–11.

153. *Parliamentary Papers*, (House of Commons), 1854 LXI 533, p.5.

154. Smith, R.A., 'What Amendments are Required in the Legislation Necessary to Prevent the Evils arising from Noxious Vapours and Smoke?' *Transactions of the National Association for the Promotion of Social Science*, 1876, pp.541–2. See also *idem*, 'Some Ancient and Modern Ideas of Sanitary Economy', *Memoirs and Proceedings of the Manchester Literary and Philosophical Society*, 2nd Series, Vol.11, 1854, pp.85–7.

155. Quoted in Smith, R.A., 'What Amendments', p.541.

156. Brüggemeier, F-J., 'A Nature Fit for Industry: The Environmental History of the Ruhr Basin, 1840–1990', *Environmental History Review*, Vol.18, 1994, pp.35–54.

157. *Ibid.*, p.37.

158. Walton, J.K., *op. cit.*, chapters 10 and 13.

159. Watkin's diary entry was made in May 1853.Quoted in Perkin, H., *The Origins of Modern English Society 1780–1880*, Routledge & Kegan Paul, London, 1972 edn, p.408.

160. Flick, C., *op. cit.*, p.36.

161. See Table 3.

162. Ransome, A., *The History of the Manchester and Salford Sanitary Association, or Half-a-Century's Progress in Sanitary Reform*, Sherratt & Hughes, Manchester, 1902.

163. *Ibid.*, p.5.

164. For example, see *Annual Report of the Manchester and Salford Sanitary Association for 1855–6*, pp.6–7. Manchester Central Library Ref., 614.06 M1.

165. The Association and its personnel are discussed in more detail in Mosley, S., 'The "Smoke Nuisance" and Environmental Reformers in Late Victorian Manchester', *Manchester Region History Review*, Vol.X, 1996, pp.40–6.

166. NVAA, *Minutes*, introductory notes.

167. *Manchester Guardian*, 3 November 1876; NVAA, *Minutes*, committee meeting, 7 July 1884.

168. NVAA, *Minutes*, committee meeting, 28 March 1877.

169. *Annual Report of the Manchester and Salford Noxious Vapours Abatement Association for 1890* (henceforth NVAA, *Annual Report*) in *Annual Report of the Manchester and Salford Sanitary Association for 1890*, p.70.

170. *Report of the Smoke Abatement Committee 1882*, Smith, Elder, & Co., London, 1883, pp.146–7.

171. See the advertisement for the Eagle Range in *ibid.*; and 'Review of Novelties, &c.', *Exhibition Review*, No.6, 1882, p.6.

172. For example, see *Manchester Guardian*, 12 December 1882.

173. *Report of the Smoke Abatement Committee 1882*, p.143; and Ranlett, J., 'The Smoke Abatement Exhibition of 1881', *History Today*, Vol.XXXI, 1981, pp.10–13.

174. *Report of the Smoke Abatement Committee 1882*, pp.6–7.

175. *Ibid.*, p.177.

176. 'Review of Novelties and Improvements', *Exhibition Review*, No.2, 1882, p.4.

177. *Report of the Smoke Abatement Committee 1882*.

178. *Ibid.*, p.144.

179. *Exhibition Review*, No.1, 1882, p.5. Also see the cartoon and article satirising the empty lecture rooms and exhibition halls and the crowded bars and bandstand at the South Kensington Sanitary Exhibition of 1884 in 'Our Insane-itary Guide to the Health Exhibition', *Punch*, 30 August 1884, p.98.

180. NVAA, *Annual Report*, 1890, p.71.

181. *Manchester Guardian*, 7 February 1883.

182. Meeting of Manchester Section, *Journal of the Society of Chemical Industry*, Vol. II, 1883, p.70.

183. *Times*, 3 August 1896.

184. NVAA, *Annual Report,* 1889, p.66.

185. Popplewell, W.C., *The Prevention of Smoke,* Scott, Greenwood & Co., London, 1901, p.136; and NVAA, *Annual Report,* 1893, p.37.

186. NVAA, *Annual Report,* 1888, Appendix A, pp.67–8.

187. NVAA, *Annual Report,* 1889, p.55; NVAA, *Annual Report,* 1888, pp.69–71; and *Manchester Guardian,* 1 December 1888.

188. Society of Chemical Industry, Manchester Section, *Minutes of General Meetings 1884–1904,* (hereafter SCI, *Minutes*), three volumes, 'The Smoke Question' 1888, microfilm, Manchester Central Library Ref., MF.1046.

189. *Ibid.*

190. *Manchester Guardian,* 1 December 1888 and 9 November 1889.

191. *Ibid.,* 20 December 1888.

192. *Ibid.,* 26 December 1888.

193. *Ibid.,* 9 November 1889.

194. *Times,* 3 August 1896.

195. *Annual Report of the Smoke Abatement League (Manchester and Salford Branch) for 1896* (hereafter SAL, *Annual Report*) in *Annual Report of the Manchester and Salford Sanitary Association for 1896,* p.40 and p.42.

196. Popplewell, W.C., *op. cit.,* pp.137–8.

197. *Ibid.,* p.139.

198. *Parliamentary Papers,* (House of Commons), 1886 (C.4845) XIV, p.12.

199. SAL (Middleton Branch), *Annual Report,* 1898, p.28.

200. *Ibid.*

201. Popplewell, W.C, *op. cit.,* p.198; and Tatham, J., *Fourteenth Annual Report on the Health of Salford,* 1883, p.33. Salford Local History Library Ref. 614. 0942. S8.

202. NVAA, *Minutes,* committee meeting, 17 June 1884.

203. *Ibid.,* committee meeting, 11 August 1881.

204. NVAA, *Annual Report,* 1894, Appendix A, pp.52–3.

205. Bryan, W.H., 'Smoke Abatement: The Rational Solution of the Problem', *Cassier's Magazine,* Vol.XIX, 1900, pp.21–2.

206. NVAA, *Annual Report,* 1894, Appendix A, pp.52–3.

207. *Ibid.,* p.46.

208. *Manchester Chamber of Commerce Monthly Record,* 29 April 1905, p.96. Microfilm, Manchester Central Library Ref., MF.1397.

209. *Ibid.,* 31 January 1906, p.7.

210. Harrison, J.F.C., *Late Victorian Britain 1875–1901,* Fontana Press, London, 1990, p.142.

211. Graham, J., 'Smoke Abatement in Manchester: Past and Present', *Sanitarian,* 1954, p.239. Moreover, London's Coal Smoke Abatement Society set up classes for the instruction of stokers at the Borough Polytechnic in 1907. See Ashby, E. and

Anderson, M., *The Politics of Clean Air*, Oxford University Press, Oxford, 1981, p.86.

212. Landes, D.S., *op. cit.*, pp.279–80; and Cardwell, D., *op. cit.*, pp.340–4.

213. *Ibid.*, p.338.

214. Landes, D.S., *op. cit.*, p.280.

215. *Report of the Smoke Abatement Committee 1882*, p.113 and pp.132–3.

216. 'Smoke Abatement', *Health Journal and Record of Sanitary Engineering*, Vol.V, 1887–8, p.190; and NVAA, *Minutes*, committee meeting, 30 August 1883.

217. Simon, S.D., *A Century of City Government, Manchester 1838–1938*, George Allen & Unwin, London, 1938, chapter XIII. For a more recent discussion see Millward, R., 'The Market Behaviour of Local Utilities in pre-World War 1 Britain: The Case of Gas', *Economic History Review*, Vol.XLIV, 1991, pp.102–27.

218. Redford, A. and Russell, I.S., *Borough*, p.306 n.

219. Smoke Abatement League of Great Britain, (Manchester and Salford Branch), *Case against the Levying of Contributions in Relief of Rates from the Profits of the Municipal Gas and Electricity Undertakings*, pamphlet, Manchester, 1912, p.8.

220. Simon, S.D., *op. cit.*, pp.362–3; and SCI, *Minutes*, 19 June 1891.

221. NVAA, *Annual Report*, 1891, Appendix C, p.91.

222. *Ibid.*, pp.90–1.

223. *Ibid.*

224. Lodge, O., 'Electricity and Gas v. Smoke', *Exhibition Review*, No.6, 1882, p.3; and Littlejohn, H., *Report on the Causes and Prevention of Smoke from Manufacturing Chimneys*, Wm. Townsend & Son, Sheffield, 1897, pp.46–7.

225. Smoke Abatement League of Great Britain, (Manchester and Salford Branch), *Case against the Levying of Contributions*, p.8.

226. *Ibid.*, Figure 1, p.5.

227. Simon, S.D., *op. cit.*, chapter XIII.

228. NVAA, *Annual Report*, 1892, p.65.

229. Church, R., *The History of the British Coal Industry, Volume 3, 1830–1913: Victorian Pre-eminence*, Clarendon Press, Oxford, 1986, Table 1.3, p.19.

230. *Ibid.*

231. *Ibid.*; and Luckin, B., *Questions of Power: Electricity and Environment in Inter-war Britain*, Manchester University Press, Manchester, 1990.

232. Frith, J., *How Electricity can Help in Abating Industrial Smoke*, Smoke Abatement League of Great Britain Pamphlet, J.F. Tangye, Manchester, 1924, pp.3–4.

233. Church, R., *op. cit.*, Table 1.3, p.19. However, some contemporaries argued that apportioning blame in this way was in itself a 'waste of energy', as both types of smoke pollution caused an 'enormous amount of harm'. See Gillespie, Rev. C.K.G., 'Some Aspects of Smoke', *Exhibition Review*, No.2, 1882, pp.2–3; and Parkes, L.C. and Des Voeux, H.A., 'Discussion on the Smoke Problem in Large Towns', *Journal of the Royal Sanitary Institute*, Vol. XXVII, 1907, p.492.

234. Horsfall, T.C., *The Nuisance of Smoke from Domestic Fires, and Methods of Abating It*, Manchester and Salford Noxious Vapours Abatement Association Pamphlet, John Heywood, Manchester, 1893, pp.3–4.

235. Arnott, N., *op. cit.*, p.430.

236. Horsfall, T.C., *op. cit.*, p.7.

237. Smoke Abatement League of Great Britain, (Manchester and Salford Branch), *Case against the levying of Contributions*, p.4.

238. See *Report of the Smoke Abatement Committee 1882*, pp.11–15 and pp.150–7.

239. Horsfall, T.C., *op. cit.*, p.15. See also *Parliamentary Papers*, (House of Commons), 1904 (Cd.2210) XXXII 145, q.5595.

240. See discussion in Arnott, N., *op. cit.*, p.433.

241. Horsfall, T.C., *op. cit.*, p.15.

242. *Cassell's Household Guide to Every Department of Practical Life*, Vol.I, Cassell Petter & Galpin, London, n.d., p.106.

243. Russell, F.A.R., *Smoke in Relation to Fogs in London*, The National Smoke Abatement Institution, London, 1889, p.32.

244. Beauchamp, J.W., *The Influence of Electricity on the Domestic Smoke Problem*, Smoke Abatement League of Great Britain Pamphlet, J.F. Tangye, Manchester, 1924, p.6.

245. NVAA, *Annual Report*, 1885, Appendix, p.62.

246. Beeton, Mrs I., *Book of Household Management*, Ward, Lock & Co., London, 1906 edn, pp.53–6; and Bone, W.A., *Coal and Health*, Pamphlet, London, 1919, p.20.

247. See Hill, L. and Campbell, A., *Health and Environment*, Edward Arnold, London, 1925, Fig.3, p.10.

248. Beeton, Mrs I., *op. cit.*, pp.54–5.

249. NVAA, *Annual Report*, 1885, Appendix, p.62.

250. NVAA, *Minutes*, committee meeting, 25 June 1885.

251. NVAA, *Minutes*, copy of a Memorial sent to the Manchester Corporation Gas Committee, 17 January 1889.

252. The number of gas appliances on hire for cooking and heating in Manchester jumped from 4,094 in 1897 to 36,214 in 1907. See Simon, S.D., *op. cit.*, p.364.

253. Bone, W.A., *op. cit.*, pp.21–2.

254. Brackenbury, C.E., *op. cit.*, p.11.

255. Long, H., *The Edwardian House: the Middle-Class Home in Britain 1880–1914*, Manchester University Press, Manchester, 1993, pp.89–90.

256. Lodge, O., *op. cit.*, p.2.

257. See *Cassell's Household Guide*, Vol.1, *op. cit.*, p.300; *Manchester Guardian*, 10 January 1889; Reid, J., 'A Case of Poisoning by Coal-Gas', *Lancet*, 27 April 1907, p.1155; and Bushell, S.M. and Gordon, C.R., 'The Housewife and Smoke Prevention', in Elliott, C. and Fitzgerald, M. (eds), *Home Fires Without Smoke: A Handbook on the Prevention of Domestic Smoke*, Ernest Benn, London, 1926, p.50.

258. Brackenbury, C.E., *op. cit.*, p.18.

259. Davidson, C., *A Woman's Work is Never Done: A History of Housework in the British Isles 1650–1950*, Chatto & Windus, London, 1982, p.99.

260. 'Cooking by Gas: Is it Injurious to the Meat?' *Lancet*, 2 February 1907, pp.306–7. See also Davidson, C., *op. cit.*, p.67.

261. The National Smoke Abatement Society, *Smoke Abatement Handbook*, p.29; and Davidson, C., *op. cit.*, pp.67–8.

262. Fishenden, M., 'Solid Fuels and Smoke Prevention', in Elliott, C. and Fitzgerald, M. (eds), *op. cit.*, p.4. See also Bushell, S.M. and Gordon, C.R., in *ibid.*, p.46.

263. Parkes, L.C. and Des Voeux, H.A., *op. cit.*, p.505.

264. Whether new appliances did actually save women time and labour in doing housework is discussed in Cowan, R.S., *More Work For Mother: The Ironies of Household Technology from the Open Hearth to the Microwave*, Basic Books, New York, 1983.

265. Shaw, Sir N. and Owens, J.S., *The Smoke Problem of Great Cities*, Constable & Co., London, 1925, p.238 and p.242.

266. Lodge, O., *op. cit.*, p.2.

267. Briggs, A., *Victorian Things*, Penguin, Harmondsworth, 1988, p.238.

268. *Parliamentary Papers*, (House of Commons), 1886 (C.4845) XIV, p.12.

269. *Manchester Guardian*, 28 May 1842; and NVAA, *Annual Report*, 1894, p.48.

270. *Ibid.*, 1893, p.37.

271. SAL (Manchester and Salford Branch), *Annual Report*, 1895, p.30.

272. NVAA, *Annual Report*, 1894, p.48.

273. See Hennock, E.P., *Fit and Proper Persons: Ideal and Reality in Nineteenth-century Urban Government*, Edward Arnold, London, 1973; and Garrard, J., *Leadership and Power in Victorian Industrial Towns 1830–80*, Manchester University Press, Manchester, 1983.

274. Kidd, A.J., 'Introduction', pp.13–15.

275. Walton, J.K., *op. cit.*, pp.222–3.

276. *Parliamentary Papers*, (House of Commons), 1904 (Cd.2210) XXXII 145, q.5582.

277. Walton, J.K., *op. cit.*, chapter 10; and Harford, I., *Manchester and its Ship Canal Movement: Class, Work and Politics in late-Victorian England*, Ryburn Publishing, Keele University Press, Keele, 1994, chapter 3.

278. *Ibid.*, p.41.

279. *Ibid.*, p.47.

280. Walton, J.K., *op. cit.*, pp.204–6. For an extended discussion see Mass, W. and Lazonick, W., 'The British Cotton Industry and International Competitive Advantage: The State of the Debates', *Business History*, Vol.23, 1990, pp.9–65.

281. Harford, I., *op. cit.*, pp.51–7.

282. 38 & 39 Vict., Ch.55, Clause 91. Emphasis added.

283. Hassan, J., *Prospects for Economic and Environmental History*, Manchester Metropolitan University, Occasional Paper, 1995, p.32.

284. Ashby, E. and Anderson, M., 'Studies in the Politics of Environmental Protection: The Historical Roots of the British Clean Air Act, 1956: III. The Ripening of Public Opinion, 1898–1952', *Interdisciplinary Science Reviews*, Vol.2, 1977, p.191.

285. Nicholson, W., *Smoke Abatement: A Manual for the Use of Manufacturers, Inspectors, Medical Officers of Health, Engineers, and Others*, Charles Griffin, London, 2nd edn, 1927, p.43; Ashby, E. and Anderson, M., *Politics*, p.87; and *Spectator*, 27 November 1926.

286. Brackenbury, C.E., *op. cit.*, p.9.

287. For the different methods smoke inspectors used in making observations, see Popplewell, W.C., *op. cit.*, pp.123–30.

288. SAL (Manchester and Salford Branch), *Annual Report*, 1898, appended table.

289. NVAA, *Annual Report*, 1884, p.76; and NVAA, *Annual Report*, 1889, Appendix E, p.71.

290. NVAA, *Minutes*, committee meetings, 21 June and 20 September 1888.

291. 25 & 26 Vict., Cap.ccv., Clause 228; and Manchester and Salford Sanitary Association, *Minute Book 1890–1907*, newspaper cutting 1897. Manchester Central Library Ref., M.126/1/1/1.

292. Gray, R., *The Factory Question and Industrial England, 1830–1860*, Cambridge University Press, Cambridge, 1996, p.171.

293. 45 & 46 Vict., Ch. cciii, Clause 44.

294. NVAA, *Annual Report*, 1884, p.76.

295. The phrase was used by Almeric W. Fitz Roy, Chairman of the Inter-Departmental Committee on Physical Deterioration, in a discussion of Manchester's smoke pollution policy. See *Parliamentary Papers*, (House of Commons), 1904 (Cd.2210) XXXII 145, q.5404.

296. Walton, J.K., *op. cit.*, p.198.

297. *Ibid.*, p.283.

298. Kidd, A.J., *Manchester*, p.103.

299. *Manchester Guardian*, 17 and 30 November 1888. See also Chubb, L.W., *Powers and Duties in the Matter of Smoke Abatement*, Coal Smoke Abatement Society Pamphlet, London, c.1906, p.7.

300. *Parliamentary Papers*, (House of Commons), 1904 (Cd.2210) XXXII 145, q.4234.

301. *Spectator*, 3 July 1926.

302. SAL (Manchester and Salford Branch), *Annual Report*, 1898, appended table.

303. *Ibid.*; and Smoke Abatement League of Great Britain, *Second Annual Report, 1910–11*, p.7.

304. *Manchester Chamber of Commerce Monthly Record*, 31 January 1895, p.6. Microfilm, Manchester Central Library Ref., MF.1395.

305. *Ibid.*, p.20.

306. Wohl, A.S., *Endangered Lives: Public Health in Victorian Britain*, Methuen, London, 1984 edn, p.223.

307. Waller, P.J., *Town, City, and Nation: England 1850–1914*, Oxford University Press, Oxford, 1983, p.280.

308. NVAA, *Annual Report*, 1894, pp.47–8.

309. SAL (Manchester and Salford Branch), *Annual Report*, 1896, p.38.

310. See *ibid.*, financial statements for the years 1895–1900.

311. SAL (Sheffield and Rotherham Branch), *Annual Report*, 1896, p.47.

312. *Ibid.*, 1897, p.33; and Nicholson, W., *op. cit.*, p.43.

313. In the United States the smoke abatement movement was actively supported by large numbers of middle class women, some even taking on the role of unofficial 'smoke inspectors'. Why fewer women were involved in British anti-smoke pressure groups is an interesting if unresolved question. For a recent discussion of women and anti-smoke activism in the United States see Stradling, D., *Smokestacks and Progressives: Environmentalists, Engineers, and Air Quality in America, 1881–1951*, Johns Hopkins University Press, Baltimore, 1999.

314. Wilkinson, T.W., 'A Crusade Against Smoke', *Humanitarian*, Vol.XVIII, 1901, pp.52–3.

315. SAL (Middleton Branch), *Annual Report*, 1900, pp.26–7. For evidence of the trifling fines imposed by the magistrates at Middleton, Rochdale and Oldham see the tables in *ibid.*, pp.28–9.

316. SAL (Sheffield and Rotherham Branch), *Annual Report*, 1898, p.31.

317. SAL (Manchester and Salford Branch), *Annual Report*, 1899, p.20.

318. NVAA, *Minutes*, report read at the committee meeting, 23 September 1881.

319. National Smoke Abatement Institution, *Report of the Meeting of the National Smoke Abatement Institution Held at the Mansion House on the 16th July 1883*, pamphlet, London, 1883, p.26.

320. SAL (Manchester and Salford Branch), *Annual Report*, 1897, p.25.

321. *Ibid.* See also Ashby, E. and Anderson, M., *Politics*, p.88.

322. *Ibid.*, p.89.

323. Pollock, W.F., 'Smoke Prevention', *Nineteenth Century*, Vol.9, 1881, p.484.

324. Thresh, J.C., *op. cit.*, p.37.

325. Horsfall, T.C., *op. cit.*, p.4.

326. *Parliamentary Papers*, (House of Commons), 1904 (Cd.2210) XXXII 145, q.5592.

327. *Ibid.*; and Horsfall, T.C., *op. cit.*, p.19.

328. Popplewell, W.C., 'A Smokeless London: The Promises of Gaseous Fuel and Electricity', *Cassier's Magazine*, Vol.XXI, 1902, p.210.

329. 'Smoke Abatement', *Health Journal and Record of Sanitary Engineering*, Vol.III, 1885–6, p.67.

330. Horsfall, T.C., *op. cit.*, pp.19–20.

331. See *Ibid.*, pp.14–15; Des Voeux, H.A., 'How to Abate the Smoke Nuisance', *Church Family Newspaper*, 20 January 1905; and Davis, G.E., 'A Piece of Coal, and What Becomes of It', *Health Journal and Record of Sanitary Engineering*, Vol.IV, 1886–7, pp.143–4.

332. Fyfe, P., 'The Pollution of the Air: Its Causes, Effects, and Cure', in Smoke Abatement League of Great Britain, *Lectures Delivered in the Technical College, Glasgow 1910–1911*, Corporation of Glasgow, Glasgow, 1912, pp.75–6.

333. Davis, G.E., *op. cit.*, p.143. See also Popplewell, W.C., 'Smokeless London', p.210.

334. Coles., W.R.E., 'Smoke Abatement', *Health Journal and Record of Sanitary Engineering*, Vol.I, 1883–4, p.90.

335. Brackenbury, C.E., *op. cit.*, p.23.

336. SCI, *Minutes*, newspaper cutting, May 1889; and 'Discussion on Town Smoke', *Journal of the Society of Chemical Industry*, Vol.XII, 1893, p.325.

337. A view shared by many of Lancashire's anti-smoke activists in the late nineteenth century. For example, see the comments of J.L. to the editor of the *Manchester Guardian*, 19 December 1888.

338. *Manchester Guardian*, 6 November 1880 and 12 December 1882.

339. Wohl, A.S., *op. cit.*, p.223 and p.218.

340. Bellamy, C., *Administering Central-Local Relations, 1871–1919: The Local Government Board in its Fiscal and Cultural Context*, Manchester University Press, Manchester, 1988, p.273.

341. Manchester and Salford Sanitary Association, *Minute Book 1890–1907*, committee meeting, 23 February 1904.

Epilogue: Too Little, Too Late?

1.   *Manchester Evening News*, 9 July 1918.

2.   Nicholson, W., *Smoke Abatement: A Manual for the Use of Manufacturers, Inspectors, Medical Officers of Health, Engineers, and Others*, Charles Griffin and Co., London, 2nd edn, 1927, p.56. First published in 1905.

3.   Ashby, E. and Anderson, M., *The Politics of Clean Air*, Oxford University Press, Oxford, 1981, p.153. See also *idem*, 'Studies in the Politics of Environmental Protection: The Historical Roots of the British Clean Air Act, 1956: I. The Awakening of Public Opinion over Industrial Smoke, 1843–53', *Interdisciplinary Science Reviews*, Vol. 1, 1976, pp.279–90; *idem*, 'Studies in the Politics of Environmental Protection: The Historical Roots of the British Clean Air Act, 1956: II. The Appeal to Public Opinion over Domestic Smoke, 1880–1892', *Interdisciplinary Science Reviews*, Vol.2, 1977, pp.9–26; and *idem*, 'Studies in the Politics of Environmental Protection: The Historical Roots of the British Clean Air Act, 1956: III. The Ripening of Public Opinion, 1898–1952', *Interdisciplinary Science Reviews*, Vol.2, 1977, pp.190–206.

4. Wohl", A.S., *Endangered Lives: Public Health in Victorian Britain*, Methuen, London, 1984 edn, p.219.

5. Bowler, C. and Brimblecombe, P., 'Control of Air Pollution in Manchester prior to the Public Health Act, 1875', *Environment and History*, Vol. 6, 2000, pp.71–98.

6. Wohl, A.S., *op. cit.*, p.219.

7. For more information on the development of Ancoats, 'the very centre of the factory system of "Cottonopolis"', consult the articles in 'Ancoats: the First Industrial Suburb', *Manchester Region History Review*, Special Issue, Vol.VII, 1993.

8. Brüggemeier, F-J., 'A Nature Fit for Industry: The Environmental History of the Ruhr Basin, 1840–1990', *Environmental History Review*, Vol. 18, 1994, p.37.

9. See Figure 2 and Szreter, S., 'The Importance of Social Intervention in Britain's Mortality Decline c.1850–1914: a Re-interpretation of the Role of Public Health', *Social History of Medicine*, Vol.1, 1988, p.13.

10. Roberts, J., '"A Densely Populated and Unlovely Tract": The Residential Development of Ancoats', *Manchester Region History Review*, Special Issue, Vol.VII, 1993, p.16.

11. Pickering, P.A., *Chartism and the Chartists in Manchester and Salford*, Macmillan, London, 1995, p.2.

12. Walton, J.K., *Chartism*, Routledge, London, 1999, p.11.

13. Clarke, A., *Moorlands and Memories, Rambles and Rides in the Fair Places of Steam-Engine Land*, Tillotsons, Bolton, 2nd edn, 1920, p.9.

14. *Manchester Guardian*, 27 June 1891.

15. Stonely, H., '"Lady" Hopwood of Middleton', *Smokeless Air*, Autumn 1952, pp.18–19.

16. *Daily News and Leader*, 28 May 1921.

17. National Smoke Abatement Society, *The London Fog: A First Survey of the December Disaster*, Pamphlet, The Leagrave Press, London, n.d., c.1953, p.6; and Ashby, E. and Anderson, M., *The Politics of Clean Air*, Oxford University Press, Oxford, pp.104–5.

18. McNeill, J.R., *Something New Under the Sun: An Environmental History of the Twentieth-century World*, Allen Lane, London, 2000, p.66.

19. See *Guardian*, 20 January 2001, p.5.

20. Dovers, S.R., 'On the Contribution of Environmental History to Current Debate and Policy', *Environment and History*, Vol.6, 2000, p.138.

21. Global Climate Coalition, 'Climate Agreement Called "Economic Disarmament"', News Release, http://www.worldcorp.com/dc-online/gcc/news/climateagreement.htm (accessed 24 July 1998). For a more recent and even gloomier assessment of the costs of implementing the Kyoto agreement see, Global Climate Coalition, Economics Committee, *The Impacts of the Kyoto Protocol*, May 2000, http://www.globalclimate.org/kyotoimpacts

22. Global Climate Coalition, 'Climate Agreement Called "Economic Disarmament"', News Release, http://www.worldcorp.com/dc-online/gcc/news/climateagreement.htm (accessed 24 July 1998).

23. See O'Connell, S., *The Car and British Society: Class, Gender and Motoring, 1896–1939*, Manchester University Press, Manchester, 1998.

24. McCarthy, T., 'The Coming Wonder? Foresight and Early Concerns about the Automobile', *Environmental History*, Vol.6, 2001, p.64.

25. Brendon, P., *The Motoring Century: The Story of the Royal Automobile Club*, Bloomsbury, London, 1997, p.380.

26. Meaton, J. and Morrice, D., 'The Ethics and Politics of Private Automobile Use', *Environmental Ethics*, Vol.18, 1996, p.39.

27. Di Chiro, G., 'Nature as Community: The Convergence of Environment and Social Justice', in Cronon, W. (ed.), *Uncommon Ground: Rethinking the Human Place in Nature*, W.W. Norton, New York, 1996, p.308. See also, McKay, G. (ed.), *DiY Culture: Party and Protest in Nineties Britain*, Verso, London, 1998; Guha, R. and Martinez-Alier, J., *Varieties of Environmentalism: Essays North and South*, Earthscan, London, 1997; and Melosi, M.V., 'Equity, Eco-Racism, and the Environmental Justice Movement', in Hughes, J.D. (ed.), *The Face of the Earth: Environment and World History*, M.E. Sharpe, New York, 2000.

28. *Transmitter. Partington and Carrington Community Newspaper*, February 1993, p.3.

29. Reclaim the Streets, 'Guerrilla Gardening. RTS Mayday Action. This is not a Protest!', http://www.gn.apc.org/rts/mayday2k/ (accessed 2 May 2000).

Bibliography

Primary Sources

Manuscript Sources

Draft Minutes of the Manchester Corporation Nuisance Committee 1856–61 & 1881–84. Manchester Central Library Refs. M.9/65/1/1–2 and M.9/65/1/8–11.

Manchester and Salford Noxious Vapours Abatement Association, *Minute Book 1876–1891*. Manchester Central Library Ref. M.126/6/1/1.

Manchester and Salford Sanitary Association, *Minute Book 1890–1907*, Manchester Central Library Ref. M.126/1/1/1.

Manchester and Salford Sanitary Association, *Minute Book 1908–1924*, Manchester Central Library Ref. M.126/1/1/2.

Society of Chemical Industry, Manchester Section, *Minutes of General Meetings 1884–1904*, three volumes, microfilm. Manchester Central Library Ref. MF.1046.

Parliamentary Records

Report from the Select Committee on Steam Engines and Furnaces, Parliamentary Papers (House of Commons), 1819 (574) VIII.

Report from the Select Committee on Steam Engines and Furnaces, Parliamentary Papers (House of Commons), 1820 (244) II.

Report from the Select Committee on Smoke Prevention, Parliamentary Papers (House of Commons), 1843 (583) VII.

First Report of the Commissioners for Inquiring into the State of Large Towns and Populous Districts, Parliamentary Papers (House of Commons), 1844 (572) XVII.I.

Second Report of the Commissioners for Inquiring into the State of Large Towns and Populous Districts, Part 1, Parliamentary Papers (House of Commons), 1845 (602) XVIII.I; *Part 2, P.P.* (House of Commons), 1845 (610) XVIII.I.

First Report from the Select Committee on Smoke Prevention, Parliamentary Papers (House of Commons), 1845 (289) XIII.

Second Report from the Select Committee on Smoke Prevention, Parliamentary Papers (House of Commons), 1845 (489) XIII.

Report by Sir H. T. De la Beche and Dr Playfair on Smoke Prohibition, Parliamentary Papers (House of Commons), 1846 (194) XLIII.

Letter by the General Board of Health with a Digest of the Information Obtained with Regard to Inventions for the Consumption of Smoke, Parliamentary Papers (House of Commons), 1854 LXI 533.

Report from the Select Committee of the House of Lords on the Smoke Nuisance Abatement (Metropolis) Bill, Parliamentary Papers (House of Lords) 1887 (321) XII.

Report of the Inter-Departmental Committee on Physical Deterioration, Vol. 1, Parliamentary Papers (House of Commons) 1904 (cd.2175) XXXII.I; *Vol.2, P.P.* (House of Commons), 1904 (cd.2210) XXXII; *Vol.3, P.P.* (House of Commons), 1904 (cd.2186) XXXII.

Annual Reports of the Chief Inspector of Alkali Works

Annual Reports of the Registrar General

Hansard, *Parliamentary Debates*

Records of Local Bodies

Annual Reports of the Manchester and Salford Noxious Vapours Abatement Association. Manchester Central Library Ref. 614.06 M1.

Annual Reports of the Smoke Abatement League (Manchester and Salford Branch). Manchester Central Library Ref. 614.06 M1.

Annual Reports of the Smoke Abatement League (Middleton Branch). Manchester Central Library Ref. 614.06 M1.

Annual Reports of the Smoke Abatement League (Sheffield and Rotherham Branch). Manchester Central Library Ref. 614.06 M1.

Annual Reports of the Smoke Abatement League of Great Britain. Archives of the National Society for Clean Air, Brighton.

Annual Reports of the Manchester and Salford Sanitary Association. Manchester Central Library Ref. 614.06 M1.

City of Manchester, Proceedings of the Council. Manchester Central Library Ref. 352.042 M22.

County Borough of Salford, Proceedings of the Council. Manchester Central Library Ref. 614.0942.S8.

Report and Proceedings of the Manchester Field-Naturalists and Archaeologists' Society, For the Year 1892, Including Second Report on the Atmosphere of Manchester and Salford, Examiner Printing Works, Manchester, 1893.

Report of the Smoke Abatement Committee 1882, Smith, Elder, & Co., London, 1883.

Newspapers, Journals and Periodicals

Bill-o'-Jack's Lancashire Monthly
British Medical Journal
Builder
Building News
Cassier's Magazine
Church Family Newspaper
Contemporary Review
Cotton Factory Times
Daily Mail
Daily News and Leader
Evening Standard and St. James Gazette
Exhibition Review
Gentleman's Magazine
Health Journal and Record of Sanitary Engineering
Household Words
Humanitarian
Illustrated London News
Journal of Agricultural Science
Journal of the Chemical Society
Journal of Gas Lighting and Water Supply
Journal of the Royal Sanitary Institute
Journal of the Society of Arts
Journal of the Society of Chemical Industry
Lancet
Manchester Courier
Manchester Chamber of Commerce Monthly Record
Manchester Evening News
Manchester Guardian
Macmillan's Magazine
Memoirs and Proceedings of the Manchester Literary and Philosophical Society
Naturalist
Nature
Nineteenth Century
Pall Mall Magazine
Philosophical Magazine

Public Health Engineer

Punch

Quarterly Journal of the Royal Meteorological Society

Quarterly Review

Sanitary Record

Spectator

Times

Times Engineering Supplement

Transactions of the Manchester Statistical Society

Transactions of the National Association for the Promotion of Social Science

Transactions of the Sanitary Institute of Great Britain

Westminster Review

Books and pamphlets

Air Pollution Advisory Board of the Manchester City Council, *The Black Smoke Tax: An Account of Damage Done by Smoke, With an Inquiry into the Comparative Cost of Family Washing in Manchester and Harrogate*, Henry Blacklock & Co., Manchester, 1920.

American Face Brick Association, *The Home Fires: A Few Suggestions in Face Brick Fireplaces*, Rogers & Co., Chicago, 1923.

Ashworth, J.R., *Smoke and the Atmosphere: Studies from a Factory Town*, Manchester University Press, Manchester, 1933.

Aston, J., *A Picture of Manchester*, 1816. Reprinted by E.J. Morten, Manchester, 1969.

Bamford, S., *Walks in South Lancashire and on its Borders*, 1844. Reprinted by Augustus M. Kelley, Clifton, New Jersey, 1972.

Banks, Mrs G.L., *The Manchester Man*, John Sherratt & Son, Altrincham, 1973 edn. First published in 1876.

Barnett, Canon S.A., and Barnett, Mrs S.A., *Practicable Socialism*, Longmans, Green & Co., London, 1915.

Baron, W., *Bits O' Broad Lancashire, Poems in the Dialect*, John Heywood, Manchester, c.1888.

Beauchamp, J.W., *The Influence of Electricity on the Domestic Smoke Problem*, Smoke Abatement League of Great Britain Pamphlet, J.F. Tangye, Manchester, 1924.

Beeton, Mrs I., *Book of Household Management*, Ward, Lock & Co., London, 1906 edn. First published in 1861.

Bennett, A., *Anna of the Five Towns*, Wordsworth Editions, Ware, 1994 edn. First published in 1902.

Black's Guide to Manchester, A. & C. Black, London, 1913.

Blaikie, W.G., *Better Days For Working People*, T. & A. Constable, Edinburgh University Press, 1881 edn.

Blatchford, R., *Merrie England*, Clarion Newspaper Company, London, 1895 edn.

Bone, W.A., *Coal and its Scientific Uses*, Longmans, Green & Co., London, 1922.

Bone, W.A., *Coal and Health*, Pamphlet, London, 1919.

Brackenbury, C.E., *Some Legal Aspects of the Smoke Nuisance*, pamphlet, King, Sell, & Olding, London, n.d., c.1910.

Bradshaw's Illustrated Guide to Manchester, Bradshaw & Blacklock, Manchester, 1857.

Brend, W.A., *Health and the State*, Constable & Co., London, 1917.

Brewer, Rev. Dr E.C., *A Guide to the Scientific Knowledge of Things Familiar*, Jarrold & Sons, London, 46th edn, 1897. First published in 1848.

Brierley, B., *'Ab - O'th' - Yate' Sketches and other Short Stories*, 3 vols., W.E. Clegg, Oldham, 1896.

Brierley, B., *Tales and Sketches of Lancashire Life*, Abel Heywood & Son, Manchester, 1885.

Brindley, W.H. (ed.), *The Soul Of Manchester*, Manchester University Press, Manchester, 1929.

Brooks, J.B., *Lancashire Bred: An Autobiography*, Church Army Press, Oxford, n.d. First published in two parts, 1950–1.

Budgen, L.M., *Live Coals; or, Faces from the Fire*, L.Reeve & Co., London, 1867.

Cassell's Household Guide to Every Department of Practical Life, 4 vols., Cassell Petter & Galpin, London, n.d. First published 1869–71.

Chadwick, E., *A Supplementary Report on the Results of a Special Inquiry into the Practice of Internment in Towns*, W. Clowes & Sons, London, 1843.

Chadwick, E., *Report on the Sanitary Condition of the Labouring Population of Great Britain*, 1842, Flinn, M.W. (ed.), Edinburgh University Press, Edinburgh, 1965.

Chalmers, A.K. (ed.), *Public Health Administration in Glasgow: A Memorial Volume of the Writings of James Burn Russell*, James Maclehose & Sons, Glasgow, 1905.

Chorley, K., *Manchester Made Them*, Faber & Faber, London, 1950.

Chubb, L.W., *Powers and Duties in the Matter of Smoke Abatement*, Coal Smoke Abatement Society Pamphlet, London, c.1906.

Chubb, L.W., *Smoke Abatement*, Coal Smoke Abatement Society Pamphlet, London, c.1912.

Clarke, A., *The Effects of the Factory System*, Grant Richards, London, 1899. George Kelsall, Littleborough, reprint of 1985.

Clarke, A., *Moorlands and Memories, Rambles and Rides in the Fair Places of Steam-Engine Land*, Tillotsons, Bolton, 2nd edn, 1920.

Coal Smoke Abatement Society, *The Coal Smoke Nuisance: Why It Should Be Abated*, Pamphlet, London, 1913.

Cohen, J.B., *The Character and Extent of Air Pollution in Leeds*, Goodall & Suddick, Leeds, 1896.

Cohen, J.B. and Ruston, A.G., *Smoke: A Study of Town Air*, Edward Arnold, London, 1912.

Coombes, B.L., *These Poor Hands: The Autobiography of a Miner Working in South Wales*, Victor Gollancz, London, 1939.

Davies, C.S., *North Country Bred: A Working-Class Family Chronicle*, Routledge & Kegan Paul, London, 1963.

Defoe, Daniel, *A Tour Through the Whole Island of Great Britain*, Vol.2, J.M. Dent & Sons, London, 1962, p. 260. First published 1724–6.

Dickens, C., *The Old Curiosity Shop*, Penguin, Harmondsworth, 1985 edn. First published in 1840–1.

Dickens, C., *A Christmas Carol: A Ghost Story of Christmas*, Penguin, Harmondsworth, 1984 edn. First published in 1843.

Dickens, C., *Christmas Books*, T.Nelson & Sons, London, n.d. First published in 1845.

Dickens, C., *Bleak House*, Bantam Books, New York, 1983 edn. First published in 1852–3.

Dickens, C., *Hard Times*, Oxford University Press, Oxford, 1989 edn. First published in 1854.

Dickens, C., *Our Mutual Friend*, Oxford University Press, Oxford, 1981 edn. First published in 1865.

Descriptive Catalogue of the Manchester Exhibition of Smoke Preventing Appliances, John Heywood, Manchester, 1882.

Disraeli, B., *Coningsby: or the New Generation*, John Lehmann, London, 1948 edn. First published in 1844.

Earwaker, J.P., *The Court Leet Records of Manchester*, 1552–1846, 12 vols., Henry Blacklock & Co., Manchester, 1884–90.

Eastlake, C.L., *Hints on Household Taste in Furniture, Upholstery and Other Details*, Longmans, Green, & Co., London, 3rd edn, 1872. Reprinted by Gregg International Publishers, Farnborough, 1971.

Elliott, C. and Fitzgerald, M. (eds), *Home Fires Without Smoke: A Handbook on the Prevention of Domestic Smoke*, Ernest Benn, London, 1926.

Encyclopaedia Britannica, Vol.V, Adam & Charles Black, Edinburgh, 9th edn, 1876.

Engels, F., *The Condition of the Working Class in England*, Penguin, Harmondsworth, 1987 edn. First published in Germany in 1845.

Evelyn, J., *Fumifugium*, W. Godbid, London, 1661. Rota, Exeter, reprint of 1976.

Faucher, L., *Manchester in 1844; Its Present Condition and Future Prospects*, Abel Heywood, Manchester, 1844. Frank Cass, London, reprint of 1969.

Fennings, A., *Fennings' Everbody's Doctor; or, When Ill, How to Get Well*, revised edn, n.d. First published in 1864.

Fitzgerald, M., *Cleansing the Sky: The Case for Smoke Abatement*, John Walker & Co., Warrington, 1924.

Fox, W., *The Working Man's Model Family Botanic Guide; or, Every Man His Own Doctor*, William Fox & Sons, Sheffield, 12th edn, 1889. First published in 1852.

Frend, W., *Is it Impossible to Free the Atmosphere of London in a Very Considerable Degree, From the Smoke and Deleterious Vapours With Which it is Hourly Impregnated?* London, 1819.

Frith, J., *How Electricity can Help in Abating Industrial Smoke*, Smoke Abatement League of Great Britain Pamphlet, J.F. Tangye, Manchester, 1924.

Froude, J.A., *Oceana, or England and her Colonies*, Longmans, Green & Co., London, 1886.

Gaskell, E., *Mary Barton: A Tale of Manchester Life*, Oxford University Press, Oxford, 1987 edn. First published in 1848.

Ginswick, J. (ed.), *Labour and the Poor in England and Wales, 1849–1851: The Letters to the Morning Chronicle from the Correspondents in the Manufacturing and Mining Districts, the Towns of Liverpool and Birmingham, and the Rural Districts, Volume 1, Lancashire, Cheshire, Yorkshire*, Frank Cass, London, 1983.

Godwin, G., *Town Swamps and Social Bridges. The Sequel of 'A Glance at the Homes of the Thousands'*, Routledge, Warnes, & Routledge, London, 1859. Leicester University Press, New York, reprint of 1972.

Graham, J.W., *The Destruction of Daylight: A Study in the Smoke Problem*, George Allen, London, 1907.

Greenwood, W., *Love on the Dole: A Tale of the Two Cities*, Jonathan Cape, London, 1935 edn. First published in 1933.

Grindon, L.H., *Manchester Walks and Wild Flowers: An Introduction to the Botany and Rural Beauty of the District*, Whittaker & Co., London, 1859.

Grindon, L.H., *The Manchester Flora: A Descriptive List of the Plants Growing Wild Within Eighteen Miles of Manchester*, William White, London, 1859.

Hamilton, H., *Scientific Treatise on Smoke Abatement*, Sherratt & Hughes, Manchester, 1917.

Harland, J. (ed.), *Lancashire Lyrics: Modern Songs and Ballads of the County Palatine*, Whittaker & Co., London, 1866.

Hartley, J., *Original Clock Almanac for 1905*, W. Nicholson & Sons, Wakefield, 1905.

Hartley, J., *Yorkshire Lyrics*, William Nicholson & Sons, London, n.d., c.1899.

Haweis, Mrs, *Rus in Urbe. or Flowers that Thrive in London Gardens and Smoky Towns*, The Leadenhall Press, London, 1886.

Hill, L. and Campbell, A., *Health and Environment*, Edward Arnold & Co., London, 1925.

Horsfall, T.C., *The Nuisance of Smoke from Domestic Fires, and Methods of Abating It*, Manchester and Salford Noxious Vapours Abatement Association Pamphlet, John Heywood, Manchester, 1893.

Horsfall, T.C., *The Place of 'Admiration, Hope, and Love' in Town Life: Reply by T.C. Horsfall to an Address presented to him on 30th June, 1910, by nearly three hundred of his Fellow-Citizens*, Pamphlet, Manchester, 1910.

Inauguration of the Albert Memorial, Manchester, R. Clark, Edinburgh, 1867.

Jevons, W.S., *The Coal Question: An Inquiry Concerning the Progress of the Nation, and the Probable Exhaustion of our Coal Mines*, Augustus M. Kelley, New York, reprint of 1965. First published in 1865.

Littlejohn, H., *Report on the Causes and Prevention of Smoke from Manufacturing Chimneys*, Wm. Townsend & Son, Sheffield, 1897.

Manchester and District Regional Smoke Abatement Committee, *Smoke Abatement and Fuel Economy in Steam Boiler Practice*, Pamphlet, Manchester, n.d., c.1921.

Manchester and Salford Sanitary Association, *Health Lectures for the People*, Vol.1, Manchester, 1878.

Manchester As It Is, Love & Barton, Manchester, 1839.

Manchester Steam Users' Association, *A Sketch of the Foundation and of the Past Fifty Year's Activity of the Manchester Steam Users' Association for the Prevention of Steam Boiler Explosions and for the Attainment of Economy in the Application of Steam*, Taylor, Garnett, Evans & Co., Manchester, 1905.

Marsh, A., *Smoke: The Problem of Coal and the Atmosphere*, Faber & Faber, London, 1947.

Mayhew, H., *London Labour and the London Poor; A Cyclopædia of the Condition and Earnings of Those that Will Work, Those that Cannot Work, and Those that Will not Work*, 4 vols., Dover Publications, New York, 1968. First published in 1861–2.

Mitchell, F.S., *The Birds of Lancashire*, Gurney & Jackson, London, 2nd edn, 1892. First published in 1885.

Morris, W. *The Collected Works of William Morris*, 24 vols, Morris, M. (ed.), Longmans Green & Co., London, 1910–15. Routledge/Thoemmes Press reprint of 1992.

Morton, H.V., *The Call of England*, Methuen, London, 4th edn, 1929.

National Smoke Abatement Institution, *Report of the Meeting of the National Smoke Abatement Institution Held at the Mansion House on the 16th July 1883*, pamphlet, London, 1883.

National Smoke Abatement Society, *The National Smoke Abatement Handbook*, The Service Guild, Manchester, 1931.

National Smoke Abatement Society, *The London Fog: A First Survey of the December Disaster*, Pamphlet, The Leagrave Press, London, n.d., c.1953.

Nicholson, W., *Smoke Abatement: A Manual for the Use of Manufacturers, Inspectors, Medical Officers of Health, Engineers, and Others*, Charles Griffin & Co., London, 2nd edn, 1927. First published in 1905.

Niven, J., *Observations on the History of Public Health Effort in Manchester*, John Heywood, Manchester, 1923.

Oliver, T. (ed.), *Dangerous Trades: The Historical, Social, and Legal Aspects of Industrial Occupations as Affecting Health, By a Number of Experts*, John Murray, London, 1902.

Ormerod, F., *Lancashire Life and Character*, Edwards & Bryning, 2nd edn, 1915.

Osborne, H., *The Problem of Atmospheric Pollution*, John Heywood, Manchester, 1924.

Pateman, T.W., *Dunshaw: A Lancashire Background*, Museum Press, London., 1948.

Penn, M., *Manchester Fourteen Miles*, Futura, London, 1982 edn. First published in 1947.

Pettigrew, W.W., *The Influence of Air Pollution on Vegetation*, Smoke Abatement League of Great Britain, Pamphlet, 1928.

Popplewell, W.C., *The Prevention of Smoke*, Scott, Greenwood & Co., London, 1901.

Procter, R.W., *Memorials of Manchester Streets*, Thos. Sutcliffe, Manchester, 1874.

Ransome, A., *The History of the Manchester and Salford Sanitary Association, or Half-a-Century's Progress in Sanitary Reform*, Sherratt & Hughes, Manchester, 1902.

Rawlinson, Sir R., *Designs for Factory, Furnace, and Other Tall Chimney Shafts*, Kell Brothers, London, 1859.

Ray, J.H. (ed.), *British Medical Association. Manchester Meeting 1902. Handbook and Guide to Manchester*, F. Ireland, Manchester, 1902.

Reach, A.B., *Manchester and the Textile Districts in 1849*, Helmshore Local History Society, reprint of 1972.

Reeves, M.P., *Round About a Pound a Week*, Virago, London, 1979. First published in 1913.

Richardson, C.J., *The Smoke Nuisance and Its Remedy. With Remarks on Liquid Fuel*, Atchley & Co., London, 1869.

Richardson, C.J., *The Englishman's House from a Cottage to a Mansion*, John Camden Hotten, London, 1870.

Roberts, R., *A Ragged Schooling: Growing up in the Classic Slum*, Manchester University Press, Manchester, 1976.

Roberts, R., *The Classic Slum: Salford Life in the First Quarter of the Century*, Penguin, Harmondsworth, 1990 edn.

Rowley, C., *Fifty Years of Work Without Wages*, The Gresham Press, London, 1912.

Ruskin, J., *The Works of John Ruskin*, Cook, E.T. and Wedderburn, A. (eds), 36 vols., George Allen, London, 1903–12.

Russell, F.A.R., *London Fogs*, Edward Stanford, London, 1880.

Russell, F.A.R., *Smoke in Relation to Fogs in London*, National Smoke Abatement Institution, London, 1889.

Russell, F.A.R., *The Atmosphere in Relation to Human Life and Health*, Smithsonian Institution, Washington, 1896.

Saleeby, C.W., *Sunlight and Health*, Nisbet & Co., London, 5th edn, 1929. First published in 1923.

Shaw, Sir N. and Owens, J.S., *The Smoke Problem of Great Cities*, Constable & Co., London, 1925.

Simon, E.D. and Fitzgerald, M., *The Smokeless City*, Longmans, Green & Co., London, 1922.

Sims, G.R. (ed.), *Living London*, vols.2 and 3, Cassell & Co., London, 1902–3.

Smith, R.A., *Air and Rain. The Beginnings of a Chemical Climatology*, Longmans, Green & Co., London, 1872.

Smoke Abatement League of Great Britain, *Proposed Branch for Manchester and District*, Pamphlet, c.1910.

Smoke Abatement League of Great Britain, *Case against the Levying of Contributions in Relief of Rates from the Profits of the Municipal Gas and Electricity Undertakings*, pamphlet, Manchester, 1912.

Smoke Abatement League of Great Britain, *Lectures Delivered in the Technical College, Glasgow 1910–1911*, Corporation of Glasgow, Glasgow, 1912.

Spence, P., *Coal, Smoke, and Sewage, Scientifically and Practically Considered*, Cave & Sever, Manchester, 1857.

Spencer, R., *A Survey of the History, Commerce and Manufactures of Lancashire*, The Biographical Publishing Company, London, 1897.

Stoker, B., *Dracula*, Oxford University Press, Oxford, 1996 edn. First published in 1897.

Sutton, C.W. (ed.), *Handbook and Guide to Manchester*, Sherratt & Hughes, Manchester, 1908.

Swindells, T., *Manchester Streets and Manchester Men*, 5 vols., J.E. Cornish, Manchester, 1906–8.

Tatham, J., *Thirteenth Annual Report on the Health of Salford, 1881*, J. Roberts, Salford, 1882.

Taylor, J.S., *Smoke and Health*, Pontefract Bros., Manchester, 1929.

Taylor, W.C., *Notes of a Tour in the Manufacturing Districts of Lancashire*, Duncan & Malcolm, London, 1842. Frank Cass, London, reprint of 1968.

The Land We Live In: A Pictorial, Historical, and Literary Sketch-Book of the British Islands, with Descriptions of their More Remarkable Features and Localities, Vol.1, William S. Orr & Co., London, n.d., c.1847.

Thresh, J.C., *An Enquiry into the Causes of the Excessive Mortality in No. 1 District, Ancoats*, John Heywood, Manchester, 1889.

Trevor, W., *Play-Times in a Busy Life: Sketches and Meditations by William Trevor*, W.E. Clegg, Oldham, 1909.

Ure, A., *The Philosophy of Manufactures: or, An Exposition of the Scientific, Moral, and Commercial Economy of the Factory System of Great Britain*, H.G. Bohn, London, 3rd edn, 1861. First published in 1835.

Vendelmans, H., *The Manual of Manures*, Country Life Library, London, 1916.

Wallin, J.E.W., *Psychological Aspects of the Problem of Atmospheric Smoke Pollution*, University of Pittsburgh, Pittsburgh, 1913.

Ward, Mrs H., *The History of David Grieve*, Smith, Elder, & Co., London, 6th edn, 1892.

Waugh, E., *The Chimney Corner*, John Heywood, Manchester, 1892. First published in 1874.

Waugh, E., *Lancashire Sketches*, Second Series, John Heywood, Manchester, 1892.

Waugh, E., *Poems and Songs*, John Heywood, Manchester, 1893.

Wheeler, J., *Manchester: Its Political, Social and Commercial History, Ancient and Modern*, Whittaker & Co., London, 1836.

Williams, C.W., *The Combustion of Coal and the Prevention of Smoke Chemically and Practically Considered*, John Weale, London, 1854.

Yates, M. (ed.), *A Lancashire Anthology*, Hodder & Stoughton, London, 1923.

Secondary Sources

Books

Alloway, B.J. and Ayres, D.C., *Chemical Principles of Environmental Pollution*, Blackie Academic & Professional, Glasgow, 1993.

Arnold, D. and Guha, R., *Nature, Culture, Imperialism*, Oxford University Press, Delhi, 1995.

Ashby, E. and Anderson, M., *The Politics of Clean Air*, Oxford University Press, Oxford, 1981.

Baldasano, J.M., Brebbia, C.A., Power, H., and Zannetti, P. (eds), *Air Pollution II Volume 1: Computer Simulation*, Computational Mechanics Publications, Southampton, 1994.

Barker, T.C. and Harris, J.R., *A Merseyside Town in the Industrial Revolution: St. Helens 1750–1900*, Frank Cass, London, 1959.

Barker, T. and Drake, M. (eds), *Population and Society in Britain 1850–1980*, Batsford Academic & Educational, London, 1982.

Beck, U., *Risk Society: Towards a New Modernity*, Sage, London, 1992.

Belchem, J., *Industrialisation and the Working Class: The English Experience, 1750–1900*, Scolar Press, Aldershot, 1991.

Bellamy, C., *Administering Central-Local Relations, 1871–1919: The Local Government Board in its Fiscal and Cultural Context*, Manchester University Press, Manchester, 1988.

Berleant, A., *The Aesthetics of Environment*, Temple University Press, Philadelphia, 1992.

Bijker, W.E. and Law, J. (eds), *Shaping Technology/Building Society: Studies in Sociotechnical Change*, MIT Press, Cambridge, Massachusetts, 1992.

Bilsky, L.J. (ed.), *Historical Ecology: Essays on Environment and Social Change*, Kennikat Press, Port Washington, New York, 1980.

Bourassa, S.C., *The Aesthetics of Landscape*, Belhaven Press, London, 1991.

Bowler, P.J., *The Fontana History of the Environmental Sciences*, Fontana Press, London, 1992.

Bradshaw, L.D. (ed.), *Visitors to Manchester: A Selection of British and Foreign Visitors' Descriptions of Manchester from c.1538 to 1865*, Neil Richardson, Swinton, 1987.

Braudel, F., *The Mediterranean and the Mediterranean World in the Age of Philip II*, Vols. I & II, Harper & Row, New York, 1972. Translation by Sîan Reynolds.

Brendon, P., *The Motoring Century: The Story of the Royal Automobile Club*, Bloomsbury, London, 1997.

Briggs, A., *Victorian Cities*, Penguin, Harmondsworth, 1990 edn.

Briggs, A., *Victorian Things*, Penguin, Harmondsworth, 1988.

Brimblecombe, P., *Air Composition and Chemistry*, Cambridge University Press, Cambridge, 2nd edn, 1996.

Brimblecombe, P., *The Big Smoke: A History of Air Pollution in London since Medieval Times*, Routledge, London, 1988.

Brimblecombe, P. and Pfister, C. (eds), *The Silent Countdown: Essays in European Environmental History*, Springer-Verlag, Berlin Heidelburg, 1990.

Brooks, M.W., *John Ruskin and Victorian Architecture*, Thames & Hudson, London, 1989.

Bullen, A. and Fowler, A., *The Cardroom Workers Union: A Centenary History of The Amalgamated Association of Card and Blowing Room Operatives*, Manchester Free Press, Manchester, 1986.

Bynum, W.F. and Porter, R. (eds), *Companion Encyclopaedia of the History of Medicine*, Volume 1, Routledge, London, 1993.

Cardwell, D., *The Fontana History of Technology*, Fontana Press, London, 1994.

Cardwell, D.S.L. (ed.), *Artisan to Graduate*, Manchester University Press, Manchester, 1974.

Carter, C.F. (ed.), *Manchester and its Region*, Manchester University Press, Manchester, 1962.

Chadwick, G.F., *The Park and the Town: Public Landscape in the 19th and 20th Centuries*, The Architectural Press, London, 1966.

Challinor, R., *The Lancashire and Cheshire Miners*, Frank Graham, Newcastle, 1972.

Chinn, C., *Poverty amidst Prosperity: The Urban Poor in England, 1834–1914*, Manchester University Press, Manchester, 1995.

Church, R., *The History of the British Coal Industry, Volume 3, 1830–1913: Victorian Pre-eminence*, Clarendon Press, Oxford, 1986.

Clapp, B.W., *An Environmental History of Britain since the Industrial Revolution*, Longman, London, 1994.

Conway, H., *People's Parks: The Design and Development of Victorian Parks in Britain*, Cambridge University Press, Cambridge, 1991.

Corbin, A., *The Foul and the Fragrant: Odour and the French Social Imagination*, Picador, London, 1994 edn. Translation by Miriam L. Kochan, Roy Porter, and Christopher Prendergast.

Corlett, W.J., *The Economic Development of Detergents*, Gerald Duckworth & Co., London, 1958.

Cowan, R.S., *More Work For Mother: The Ironies of Household Technology from the Open Hearth to the Microwave*, Basic Books, New York, 1983.

Cronon, W., *Nature's Metropolis: Chicago and the Great West*, W.W. Norton, New York, 1992 edn.

Cronon, W. (ed.), *Uncommon Ground: Rethinking the Human Place in Nature*, W.W. Norton, New York, 1996.

Cullen, M.J., *The Statistical Movement in Early Victorian Britain: The Foundations of Empirical Social Research*, The Harvester Press, Hassocks, 1975.

Davidson, C., *A Woman's Work is Never Done: A History of Housework in the British Isles 1650–1950*, Chatto & Windus, London, 1982.

Davies, A. and Fielding, S. (eds), *Workers' Worlds: Cultures and Communities in Manchester and Salford, 1880–1939*, Manchester University Press, Manchester, 1992.

Diggle, G.E., *A History of Widnes*, Wilmer Bros. & Haram, Birkenhead, 1961.

Douglas, M., *Purity and Danger: An Analysis of Concepts of Pollution and Taboo*, Routledge & Kegan Paul, London, 1966.

Douglas, M., *Implicit Meanings: Essays in Anthropology*, Routledge & Kegan Paul, London, 1975.

Douglas, M. and Wildavsky, A., *Risk and Culture: An Essay on the Selection of Technical and Environmental Dangers*, University of California Press, Berkeley, 1982.

Douglas, M., *Risk and Blame: Essays in Cultural Theory*, Routledge, London, 1992.

Douglas, M., *Thought Styles: Critical Essays on Good Taste*, Sage, London, 1996.

Dutton, R., *The Victorian Home: Some Aspects of Nineteenth-century Taste and Manners*, B.T. Batsford, London, 1954.

Dyos, H.J. and Wolff, M. (eds), *The Victorian City*, 2 vols., Routledge & Kegan Paul, London, 1973.

Evans, R.J., *Death in Hamburg: Society and Politics in the Cholera Years 1830–1910*, Penguin, Harmondsworth, 1990 edn.

Farnie, D., *The English Cotton Industry and the World Market, 1815–96*, Clarendon Press, Oxford, 1979.

Flinn, M.W., *The History of the British Coal Industry, Volume 2, 1700–1830: The Industrial Revolution*, Clarendon Press, Oxford, 1984, p.231.

Fowler, A. and Wyke, T. (eds), *The Barefoot Aristocrats: A History of the Amalgamated Association of Cotton Spinners*, George Kelsall, Littleborough, 1987.

Fraser, D., *The Evolution of the British Welfare State: A History of Social Policy since the Industrial Revolution*, Macmillan Education, London, 1992 edn.

Fraser, D. (ed.), *Municipal Reform in the Industrial City*, Leicester University Press, Leicester, 1982.

Fraser, D. and Sutcliffe, A. (eds), *The Pursuit of Urban History*, Edward Arnold, London, 1983.

Garrard, J., *Leadership and Power in Victorian Industrial Towns 1830–80*, Manchester University Press, Manchester, 1983.

Gloag, J., *The Englishman's Castle: A History of Houses, Large and Small, in Town and Country, from A.D. 100 to the Present Day*, Eyre & Spottiswoode, London, 1944.

Gould, P.C., *Early Green Politics: Back to Nature, Back to the Land, and Socialism in Britain, 1880–1900*, Harvester Press, Brighton, 1988.

Gray, R., *The Factory Question and Industrial England, 1830–1860*, Cambridge University Press, Cambridge, 1996.

Griffiths, T. and Robin, L. (eds), *Ecology and Empire: Environmental History of Settler Societies*, Keele University Press, Edinburgh, 1997.

Guha, R. and Martinez-Alier, J., *Varieties of Environmentalism: Essays North and South*, Earthscan, London, 1997.

Hajer, M.A., *The Politics of Environmental Discourse: Ecological Modernisation and the Policy Process*, Clarendon Press, Oxford, 1995.

Hamlin, C., *What Becomes of Pollution? Adversary Science and the Controversy on the Self-Purification of Rivers in Britain, 1850–1900*, Garland Publishing, New York, 1987.

Hamlin, C., *Public Health and Social Justice in the Age of Chadwick: Britain, 1800–1854*, Cambridge University Press, Cambridge, 1998.

Hardy, A., *The Epidemic Streets: Infectious Disease and the Rise of Preventive Medicine, 1856–1900*, Clarendon Press, Oxford, 1993.

Harford, I., *Manchester and its Ship Canal Movement: Class, Work and Politics in Late-Victorian England*, Ryburn Publishing, Keele University Press, Keele, 1994.

Harrison, J.F.C., *Late Victorian Britain, 1875–1901*, Fontana Press, London, 1990.

Hassan, J.A., *A History of Water in Modern England and Wales*, Manchester University Press, Manchester, 1998.

Hatcher, J., *The History of the British Coal Industry, Volume 1, Before 1700: Towards the Age of Coal*, Clarendon Press, Oxford, 1993.

Hennock, E.P., *Fit and Proper Persons: Ideal and Reality in Nineteenth-century Urban Government*, Edward Arnold, London, 1973.

Hobsbawm, E.J., *Labouring Men: Studies in the History of Labour*, Weidenfield & Nicolson, London, 1964.

Horsfield, M., *Biting the Dust: The Joys of Housework*, Fourth Estate, London, 1997.

Hoy, S., *Chasing Dirt: The American Pursuit of Cleanliness*, Oxford University Press, New York, 1995.

Hughes, J.D. (ed.), *The Face of the Earth: Environment and World History*, M.E. Sharpe, New York, 2000.

Jaritz, G. and Winiwarter, V. (eds), *Umweltbewältigung: Die Historische Perspektive*, Verlag Für Regionalgeschichte, Bielefeld, 1994.

Kargon, R.H., *Science in Victorian Manchester: Enterprise and Expertise*, Manchester University Press, Manchester, 1977.

Kearns, G. and Withers, C.W.J. (eds), *Urbanising Britain: Essays on Class and Community in the Nineteenth Century*, Cambridge University Press, Cambridge, 1991.

Kennedy, P., *The Rise and Fall of the Great Powers: Economic Change and Military Conflict from 1500 to 2000*, Fontana Press, London, 1988.

Kidd, A.J. and Roberts, K.W. (eds), *City, Class and Culture: Studies of Social Policy and Cultural Production in Victorian Manchester*, Manchester University Press, Manchester, 1985.

Kidd, A.J., *Manchester*, Ryburn Publishing, Keele University Press, Keele, 1993.

Kidd, A.J. and Nicholls, D. (eds), *Gender, Civic Culture and Consumerism: Middle-Class Identity in Britain, 1800–1940*, Manchester University Press, Manchester, 1999.

Landes, D.S., *The Unbound Prometheus: Technological Change and Industrial Development in Western Europe from 1750 to the Present*, Cambridge University Press, New York, 1969.

Langton, J., *Geographical Change and Industrial Revolution: Coalmining in South West Lancashire, 1590–1799*, Cambridge University Press, Cambridge, 1979.

Langton, J. and Morris, R.J. (eds), *Atlas of Industrialising Britain, 1780–1914*, Methuen & Co., London, 1986.

Lawton, R. and Pooley, C.G., *Britain 1740–1950: An Historical Geography*, Edward Arnold, London, 1992.

Lichatowich, J., *Salmon Without Rivers: A History of the Pacific Salmon Crisis*, Island Press, Washington D.C., 1999.

Lloyd-Jones, R. and Lewis, M.J., *Manchester and the Age of the Factory: The Business Structure of Cottonopolis in the Industrial Revolution*, Croom Helm, Beckenham, 1988.

Long, H., *The Edwardian House: the Middle-Class Home in Britain 1880–1914*, Manchester University Press, Manchester, 1993.

Longhurst, J.W.S. (ed.), *Acid Deposition: Sources, Effects and Controls*, British Library and Technical Communications, Information Press, Oxford, 1989.

Luckin, B., *Pollution and Control: A Social History of the Thames in the Nineteenth Century*, Adam Hilger, Bristol, 1986.

Luckin, B., *Questions of Power: Electricity and Environment in Inter-war Britain*, Manchester University Press, Manchester, 1990.

MacLeod, R. (ed.), *Government and Expertise: Specialists, Administrators and Professionals, 1860–1919*, Cambridge University Press, Cambridge, 1988.

Maidment, B., *The Poorhouse Fugitives: Self-taught Poets and Poetry in Victorian Britain*, Carcanet Press, Manchester, 1992 edn.

Manley, G., *Climate and the British Scene*, Collins, London, 1971 edn.

Marcus, S., *Engels, Manchester, and the Working Class*, Weidenfeld & Nicolson, London, 1974.

McKay, G. (ed.), *DiY Culture: Party and Protest in Nineties Britain*, Verso, London, 1998.

McNeill, J.R., *Something New Under the Sun: An Environmental History of the Twentieth-century World*, Allen Lane, London, 2000.

Meadows, D.H., Meadows, D.L., and Randers, J., *Beyond the Limits: Global Collapse or a Sustainable Future*, Earthscan, London, 1995 edn.

Melosi, M.V. (ed.), *Pollution and Reform in American Cities, 1870–1930*, University of Texas Press, Austin, 1980.

Melosi, M.V., *The Sanitary City: Urban Infrastructure in America from Colonial Times to the Present*, Johns Hopkins University Press, Baltimore, 2000.

Merleau-Ponty, M., *Phenomenology of Perception*, Routledge, London, 1996 edn. Translation by Colin Smith.

Midwinter, E.C., *Social Administration in Lancashire, 1830–1860. Poor Law, Public Health, Police*, Manchester University Press, Manchester, 1969.

Milton, K. (ed.), *Environmentalism: The View from Anthropology*, Routledge, London, 1993.

Milton, K., *Environmentalism and Cultural Theory: Exploring the Role of Anthropology in Environmental Discourse*, Routledge, London, 1996.

Mohun, A.P., *Steam Laundries: Gender, Technology, and Work in the United States and Great Britain*, Johns Hopkins University Press, Baltimore, 1999.

Morris, R.J., *Cholera 1832: The Social Response to an Epidemic*, Croom Helm, London, 1976.

Mumford, L., *The City in History: Its Origins, Its Transformations, and Its Prospects*, Secker & Warburg, London, 1961.

Musson, A.E., *Enterprise in Soap and Chemicals: Joseph Crosfield and Sons, Ltd. 1815–1965*, Manchester University Press, Manchester, 1965.

Musson, A.E. and Robinson, E., *Science and Technology in the Industrial Revolution*, Manchester University Press, Manchester, 1969.

Nef, J.U., *The Rise of the British Coal Industry*, Vol.1, George Routledge & Sons, London, 1932.

Nye, D.E., *Electrifying America: Social Meanings of a New Technology, 1880–1940*, MIT Press, Cambridge, Massachusetts, 1990.

Nye, D.E., *Narratives and Spaces: Technology and the Construction of American Culture*, University of Exeter Press, Exeter, 1997.

Oakes, C., *The Birds of Lancashire*, Oliver & Boyd, Edinburgh, 1953.

O'Connell, S., *The Car and British Society: Class, Gender and Motoring, 1896–1939*, Manchester University Press, Manchester, 1998.

Offer, A., *Property and Politics 1870–1914: Landownership, Law, Ideology and Urban Development in England*, Cambridge University Press, Cambridge, 1981.

Pepper, D., *Modern Environmentalism: An Introduction*, Routledge, London, 1996.

Perkin, H., *The Origins of Modern English Society 1780–1880*, Routledge & Kegan Paul, London, 1972 edn.

Pickering, P.A., *Chartism and the Chartists in Manchester and Salford*, Macmillan, London, 1995.

Porter, B., *The Lion's Share: A Short History of British Imperialism 1850–1983*, Longman, Harlow, 2nd edn, 1984.

Porter, D. (ed.), *The History of Public Health and the Modern State*, Clio Medica 26 / The Wellcome Institute Series in the History of Medicine, Rodopi B.V., Amsterdam, 1994.

Porter, D., *The Thames Embankment: Environment, Technology, and Society in Victorian London*, University of Akron Press, Akron, 1998.

Potter, J. and Wetherell, M., *Discourse and Social Psychology: Beyond Attitudes and Behaviour*, Sage, London, 1994 edn.

Praz, M., *An Illustrated History of Interior Decoration from Pompeii to Art Nouveau*, Thames & Hudson, London, 1964.

Ratcliffe, D.A. (ed.), *A Nature Conservation Review*, Vol.1, Cambridge University Press, Cambridge, 1977.

Ravetz, A., with Turkington, R., *The Place of Home: English Domestic Environments, 1914–2000*, E. & F.N. Spon, London, 1995.

Read, C. (ed.), *How Vehicle Pollution Affects Our Health*, Abacus Printing Co., London, 1994.

Redford, A. and Russell, I.S., *The History of Local Government in Manchester. Volume I: Manor and Township*, Longmans, Green & Co., London, 1939.

Redford, A. and Russell, I.S., *The History of Local Government in Manchester. Volume II: Borough and City*, Longmans, Green & Co., London, 1940.

Rees, R., *King Copper: South Wales and the Copper Trade*, University of Wales Press, Cardiff, 2000.

Richardson, D.H.S., *Pollution Monitoring with Lichens*, Richmond Publishing, Slough, 1992.

Roberts, E., *A Woman's Place: An Oral History of Working-Class Women 1890–1940*, Basil Blackwell, London, 1985 edn.

Rose, M.B., *The Gregs of Quarry Bank Mill: The Rise and Decline of a Family Firm, 1750–1914*, Cambridge University Press, Cambridge, 1986.

Rose, M.B. (ed.), *The Lancashire Cotton Industry: A History Since 1700*, Lancashire County Books, Preston, 1996.

Ross, E., *Love and Toil: Motherhood in Outcast London, 1870–1918*, Oxford University Press, New York, 1993.

Royle, E., *Chartism*, Longman, London, 1980.

Scola, R., *Feeding the Victorian City: The food supply of Manchester, 1770–1870*, Manchester University Press, Manchester, 1992.

Schutz, A., *On Phenomenology and Social Relations: Selected Writings*, edited by Helmut R. Wagner, University of Chicago Press, Chicago, 1970.

Bibliography

Shepard, P., *Man in the Landscape: A Historic View of the Esthetics of Nature*, A&M University Press, College Station, Texas, 1991 edn.

Sigsworth, E.M. (ed.), *In Search of Victorian Values: Aspects of Nineteenth-century Thought and Society*, Manchester University Press, Manchester, 1988.

Simon, S.D., *A Century of City Government, Manchester 1838–1938*, George Allen & Unwin, London, 1938.

Smith, C. and Wise, M.N., *Energy and Empire: A Biographical Study of Lord Kelvin*, Cambridge University Press, Cambridge, 1989.

Smith, F.B., *The People's Health 1830–1910*, Croom Helm, London, 1979.

Soulé, M.E. and Lease, G. (eds), *Reinventing Nature? Responses to Postmodern Deconstruction*, Island Press, Washington, 1995.

Stewart, C., *The Stones of Manchester*, Edward Arnold, London, 1956.

Stradling, D., *Smokestacks and Progressives: Environmentalists, Engineers, and Air Quality in America, 1881–1951*, Johns Hopkins University Press, Baltimore, 1999.

Stratton, M., *The Terracotta Revival: Building Innovation and the Image of the Industrial City in Britain and North America*, Victor Gollancz, London, 1993.

Sutcliffe, A. (ed.), *British Town Planning: the Formative Years*, Leicester University Press, Leicester, 1981.

The Village Atlas: The Growth of Manchester, Lancashire & North Cheshire, 1840–1912, The Alderman Press/The Village Press, Edmonton, 1989.

Tarr, J.A., *The Search for the Ultimate Sink: Urban Pollution in Historical Perspective*, University of Akron Press, Akron, 1996.

Thomas, K., *Man and the Natural World: Changing Attitudes in England 1500–1800*, Penguin, Harmondsworth, 1984 edn.

Thompson, D., *The Chartists: Popular Politics in the Industrial Revolution*, Wildwood House, Aldershot, 1986.

Thompson, F.M.L., *The Rise of Respectable Society: A Social History of Victorian Britain, 1830–1900*, Fontana Press, London, 1988.

Treblicock, C., *The Industrialisation of the Continental Powers, 1780–1914*, Longman, Harlow, 1989 edn.

Tuan, Y-F., *Topophilia: A Study of Environmental Perception, Attitudes, and Values*, Columbia University Press, New York, Morningside edn, 1990.

Vigier, F., *Change and Apathy: Liverpool and Manchester during the Industrial Revolution*, MIT Press, Cambridge, Massachusetts, 1970.

Ward, J.T. (ed.), *Popular Movements c.1830–1850*, Macmillan & Co., London, 1970.

Waller, P.J., *Town, City, and Nation: England 1850–1914*, Oxford University Press, Oxford, 1983.

Walton, J.K., *Lancashire: A Social History, 1558–1939*, Manchester University Press, Manchester, 1990 edn.

Walton, J.K., *Chartism*, Routledge, London, 1999.

Webb, S. and Webb, B., *English Local Government from the Revolution to the Municipal Corporations Act: The Manor and the Borough*, part 1, Longmans, Green & Co., London, 1908.

Wellburn, A., *Air Pollution and Climate Change: The Biological Impact*, Longman Scientific & Technical, Harlow, 2nd edn, 1994.

Wheeler, M. (ed.), *Ruskin and Environment: The Storm-cloud of the Nineteenth Century*, Manchester University Press, Manchester, 1995.

Williams, R., *Keywords: A Vocabulary of Culture and Society*, Flamingo, London, 1976.

Wilmer, C. (ed.), *Unto This Last, and Other Writings by John Ruskin*, Penguin, Harmondsworth, 1985.

Wohl, A.S., *Endangered Lives: Public Health in Victorian Britain*, Methuen & Co., London, 1984 edn.

Wolff, J. and Seed, J. (eds), *The Culture of Capital: Art, Power and the Nineteenth-century Middle Class*, Manchester University Press, Manchester, 1988.

Wood, C.M., Lee, N., Luker, J.A., and Saunders, P.J.W., *The Geography of Pollution: A Study of Greater Manchester*, Manchester University Press, Manchester, 1974.

Woods, R. and Woodward, J. (eds), *Urban Disease and Mortality in Nineteenth-century England*, Batsford Academic & Educational, London, 1984.

World Commission for Environment and Development, *Our Common Future*, The Brundtland Report, Oxford University Press, Oxford, 1987.

Worster, D. (ed.), *The Ends of the Earth: Perspectives on Modern Environmental History*, Cambridge University Press, New York, 1988.

Worster, D., *Nature's Economy: A History of Ecological Ideas*, Cambridge University Press, New York, 2nd edn, 1994.

Articles

Ashby, E. and Anderson, M., 'Studies in the Politics of Environmental Protection: The Historical Roots of the British Clean Air Act, 1956: I. The Awakening of Public Opinion over Industrial Smoke, 1843–53', *Interdisciplinary Science Reviews*, Vol.1, 1976, pp.279–90.

Ashby, E. and Anderson, M., 'Studies in the Politics of Environmental Protection: The Historical Roots of the British Clean Air Act, 1956: II. The Appeal to Public Opinion over Domestic Smoke, 1880–1892', *Interdisciplinary Science Reviews*, Vol.2, 1977, pp.9–26.

Ashby, E. and Anderson, M., 'Studies in the Politics of Environmental Protection: The Historical Roots of the British Clean Air Act, 1956: III. The Ripening of Public Opinion, 1898–1952', *Interdisciplinary Science Reviews*, Vol.2, 1977, pp.190–206.

Battarbee, R.W., 'Diatom Analysis and the Acidification of Lakes', *Philosophical Transactions of the Royal Society of London*, Series B, Vol.305, 1984, pp.451–77.

Beck, A., 'Some Aspects of the History of Anti-Pollution Legislation in England, 1819–1954', *Journal of the History of Medicine and Allied Sciences*, Vol.14, 1959, pp.475–89.

Bess, M., Cioc, M., and Sievert, J., 'Environmental History Writing in Southern Europe', *Environmental History*, Vol.5, 2000, pp.545–6.

Bird, E.A.R., 'The Social Construction of Nature: Theoretical Approaches to the History of Environmental Problems', *Environmental Review*, Vol.11, 1987, pp.255–64.

Bowler, C. and Brimblecombe, P., 'Control of Air Pollution in Manchester prior to the Public Health Act, 1875', *Environment and History*, Vol.6, 2000, pp.71–98.

Brenner, J.F., 'Nuisance Law and the Industrial Revolution', *Journal of Legal Studies*, Vol.III, 1974, pp.403–33.

Brüggemeier, F-J., 'A Nature Fit for Industry: The Environmental History of the Ruhr Basin, 1840–1990', *Environmental History Review*, Vol.18, 1994, pp.35–54.

Chase, M., 'Can History be Green? A Prognosis', *Rural History*, Vol.3, 1992, pp.243–51.

Cherfas, J., 'Clean Air Revives the Peppered Moth', *New Scientist*, Vol.109, 1986, p.17.

Cioc, M., Linner, B-O., and Osborn, M., 'Environmental History Writing in Northern Europe', *Environmental History*, Vol.5, 2000, pp.396–406.

Coates, P., 'Clio's New Greenhouse', *History Today*, Vol.46, 1996, pp.15–22.

Constantine, S., 'Amateur Gardening and Popular Recreation in the 19th and 20th Centuries', *Journal of Social History*, Vol.14, 1980–81, pp.387–406.

Cook, L.M., Askew, R.R., and Bishop, J.A., 'Increasing Frequency of the Typical Form of the Peppered Moth in Manchester', *Nature*, Vol.227, 1970, p.1155.

Cowling, E.B., 'Acid Precipitation in Historical Perspective', *Environmental Science and Technology*, Vol.16, 1982, pp.110A–23A.

Cronon, W., 'A Place for Stories: Nature, History, and Narrative', *Journal of American History*, Vol.78, 1992, pp.1347–76.

Cronon, W., 'Comment: Cutting Loose or Running Aground?' *Journal of Historical Geography*, Vol.20, 1994, pp.38–43.

Crosby, A.W., 'The Past and Present of Environmental History', *American Historical Review*, Vol.100, 1995, pp.1177–89.

Cumbler, J.T., 'Whatever Happened to Industrial Waste?: Reform, Compromise, and Science in Nineteenth Century Southern New England', *Journal of Social History*, Vol.29, 1995, pp.149–71.

Demeritt, D., 'Ecology, Objectivity and Critique in Writings on Nature and Human Societies', *Journal of Historical Geography*, Vol.20, 1994, pp.22–37.

Dingle, A.E, '"The Monster Nuisance of All": Landowners, Alkali Manufacturers, and Air Pollution, 1828–64', *Economic History Review*, Second Series, Vol.XXXV, 1982, pp.529–48.

Dovers, S.R., 'On the Contribution of Environmental History to Current Debate and Policy', *Environment and History*, Vol.6, 2000, pp.131–50.

Eden, S., 'Public Participation in Environmental Policy: Considering Scientific, Counter-scientific and Non-scientific Contributions', *Public Understanding of Science*, Vol.5, 1996, pp.183–204.

Eddy, T.P., 'Coal Smoke and Mortality of the Elderly', *Nature*, Vol.251, 1974, pp.136–38.

Elesh, D., 'The Manchester Statistical Society: A Case Study of a Discontinuity in the History of Empirical Social Research', Parts 1 and 2, *Journal of the History of the Behavioral Sciences*, Vol.8, 1972, pp.280–301 and pp.407–17.

Ferguson, P. and Lee, J.A., 'Past and Present Sulphur Pollution in the Southern Pennines', *Atmospheric Environment*, Vol.17, 1983, pp.1131–37.

Flick, C., 'The Movement for Smoke Abatement in 19th-Century Britain', *Technology and Culture*, Vol.21, 1980, pp.29–50.

Frankel, M., 'The Alkali Inspectorate: The Control of Industrial Air Pollution', *Social Audit*, Vol.1, 1974, pp.3–48.

Gaskell, S.M., 'Gardens for the Working Class: Victorian Practical Pleasure', *Victorian Studies*, Vol.23, 1980, pp.479–501.

Gatrell, V.A.C., 'Labour, Power, and the Size of Firms in Lancashire Cotton in the Second Quarter of the Nineteenth Century', *Economic History Review*, 2nd Series, Vol.XXX, 1977, pp.95–129.

Gibson, A. and Farrar, W.V., 'Robert Angus Smith, F.R.S., and "Sanitary Science"', *Notes and Records of the Royal Society of London*, Vol.28, 1973–4, pp.241–62.

Goddard, N., '"A mine of wealth"? The Victorians and the Agricultural Value of Sewage', *Journal of Historical Geography*, Vol.22, 1996, pp.274–90.

Graham, J., 'Smoke Abatement in Manchester: Past and Present', *Sanitarian*, 1954, pp.235–9.

Greider, T. and Garkovich, L., 'Landscapes: The Social Construction of Nature and the Environment', *Rural Sociology*, Vol.59, 1994, pp.1–24.

Hamlin, C., 'Providence and Putrefaction: Victorian Sanitarians and the Natural Theology of Health and Disease', *Victorian Studies*, Vol.28, 1985, pp.381–411.

Hamlin, C., 'Muddling in Bumbledom: On the Enormity of Large Sanitary Improvements in Four British Towns', *Victorian Studies*, Vol.32, 1988, pp.55–83.

Hamlin, C., 'Edwin Chadwick and the Engineers, 1842–1854: Systems and Antisystems in the Pipe-and-Brick Sewers War', *Technology and Culture*, Vol.33, 1992, pp.680–709.

Hamlin, C., 'Environmental Sensibility in Edinburgh, 1839–1840: The 'Fetid Irrigation' Controversy', *Journal of Urban History*, Vol.20, 1994, pp.311–39.

Hardy, A., 'Urban Famine or Urban Crisis? Typhus in the Victorian City', *Medical History*, Vol.32, 1988, pp.401–25.

Hardy, A., 'Rickets and the Rest: Child-care, Diet and the Infectious Children's Diseases, 1850–1914', *Social History of Medicine*, Vol.5., 1992, pp.389–412.

Harrison, B., 'Philanthropy and the Victorians', *Victorian Studies*, Vol.IX, 1966, pp.353–74.

Harrison, M., 'Art and Social Regeneration: The Ancoats Art Museum', *Manchester Region History Review*, Vol.VII, 1993, pp.63–72.

Hassan, J.A., 'The Impact and Development of the Water Supply in Manchester, 1568–1882', *Transactions of the Historic Society of Lancashire and Cheshire*, Vol.133, 1984, pp.25–45.

Hassan, J.A., 'The Growth and Impact of the British Water Industry in the Nineteenth Century', *Economic History Review*, 2nd Series, Vol.XXXVIII, 1985, pp.531–47.

Hassan, J.A., *Prospects for Economic and Environmental History*, Manchester Metropolitan University, Occasional Paper, 1995.

Hawes, R., 'The Control of Alkali Pollution in St Helens, 1862–1890', *Environment and History*, Vol.1, 1995, pp.159–71.

Hewitt, M., 'The Travails of Domestic Visiting: Manchester, 1830–70', *Historical Research*, Vol.71, 1998, pp.196–227.

Hitchcock, H-R., 'Victorian Monuments of Commerce', *Architectural Review*, Vol.105, 1949, pp.61–74.

Horsman, D.C., Roberts, T.M., and Bradshaw, A.D., 'Evolution of Sulphur Dioxide Tolerance in Perennial Ryegrass', *Nature*, Vol.276, 1978, p.493.

Jenner, M., 'The Politics of London Air: John Evelyn's *Fumifugium* and the Restoration', *Historical Journal*, Vol.38, 1995, pp.535–51.

Kay, A., 'Charles Rowley and the Ancoats Recreation Movement, 1876–1914', *Manchester Region History Review*, Vol.VII, 1993, pp.45–54.

Kearns, G., 'Private Property and Public Health Reform in England 1830–70', *Social Science and Medicine*, Vol.26, 1988, pp.187–99.

Kidd, A.J., 'The Industrial City and its Pre-industrial Past: the Manchester Royal Jubilee Exhibition of 1887', *Transactions of the Lancashire and Cheshire Antiquarian Society*, Vol.89, 1993, pp.54–73.

Loomis, W.F., 'Rickets', *Scientific American*, Vol.223, 1970, pp.76–91.

Lowe, J., 'In Praise of Tall Chimneys', *Industrial Heritage*, Vol.8, 1989, pp.11–19.

Luckin, B., 'Death and Survival in the City: Approaches to the History of Disease', *Urban History Yearbook*, 1980, pp.53–62.

Luckin, B. and Mooney, G., 'Urban History and Historical Epidemiology: The Case of London, 1860–1920', *Urban History*, Vol.24, 1997, pp.37–55.

McCarthy, T., 'The Coming Wonder? Foresight and Early Concerns about the Automobile', *Environmental History*, Vol.6, 2001, pp.46–74.

McLaren, J.P.S., 'Nuisance Law and the Industrial Revolution – Some Lessons from Social History', *Oxford Journal of Legal Studies*, Vol.3, 1983, pp.155–221.

Macleod, R.M., 'The Alkali Acts Administration, 1863–84: The Emergence of the Civil Scientist', *Victorian Studies*, Vol.IX, 1965, pp.85–112.

Malcolm, C.V., 'Smokeless Zones – The History of Their Development', parts 1 and 2, *Clean Air*, Autumn 1976 and Spring 1977, pp.14–20 and pp.4–10.

Mass, W. and Lazonick, W., 'The British Cotton Industry and International Competitive Advantage: The State of the Debates', *Business History*, Vol.23, 1990, pp.9–65.

Meaton, J. and Morrice, D., 'The Ethics and Politics of Private Automobile Use', *Environmental Ethics*, Vol.18, 1996, pp.39–54.

Melosi, M.V., 'The Place of the City in Environmental History', *Environmental History Review*, Vol.17, 1993, pp.1–23.

Merricks, L., 'Environmental History', *Rural History*, Vol.7, 1996, pp.97–109.

Millward, R., 'The Market Behaviour of Local Utilities in pre-World War 1 Britain: The Case of Gas', *Economic History Review*, Vol.XLIV, 1991, pp.102–27.

Morris, R.J., 'Voluntary Societies and British Urban Elites, 1780–1850: An Analysis', *Historical Journal*, Vol.26, 1983, pp.95–118.

Mosley, S., 'The "Smoke Nuisance" and Environmental Reformers in Late Victorian Manchester', *Manchester Region History Review*, Vol.X, 1996, pp.40–6.

Musson, A.E., 'Industrial Motive Power in the United Kingdom, 1800–70', *Economic History Review*, 2nd Series, Vol.XXIX, 1976, pp.415–39.

Newell, E., 'Atmospheric Pollution and the British Copper Industry, 1690–1920', *Technology and Culture*, Vol.38, 1997, pp.655–89.

O'Riordan, T., 'Culture and the Environment in Britain', *Environmental Management*, Vol.9, 1985, pp.113–120.

Pass, A., 'Thomas Worthington: Practical Idealist', *Architectural Review*, Vol.155, 1974, pp.268–74.

Platt, H.L., 'Invisible Gases: Smoke, Gender, and the Redefinition of Environmental Policy in Chicago, 1900–1920', *Planning Perspectives*, Vol.10, 1995, pp.67–97.

Platt, H.L., 'The Emergence of Urban Environmental History', *Urban History*, Vol.26, 1999, pp.89–95.

Pollard, A., '"Sooty Manchester" and the Social Reform Novel', *Journal of Industrial Medicine*, Vol.18, 1961, pp.85–92.

Pollard, S., 'A New Estimate of British Coal Production', *Economic History Review*, 2nd Series, Vol.XXXIII, 1980, pp.212–235.

Porter, D., '"Enemies of the Race": Biologism, Environmentalism, and Public Health in Edwardian England', *Victorian Studies*, Vol.34, 1991, pp.159–178.

Ranlett, J., 'The Smoke Abatement Exhibition of 1881', *History Today*, Vol.XXXI, 1981, pp.10–13.

Ranlett, J., '"Checking Nature's Desecration": Late-Victorian Environmental Organisation', *Victorian Studies*, Vol.26, 1983, pp.197–222.

Ravetz, A., 'The Victorian Coal Kitchen and its Reformers', *Victorian Studies*, Vol.XI, 1968, pp.435–60.

Rees, R., 'The South Wales Copper-Smoke Dispute, 1833–95', *Welsh History Review*, Vol.10, 1980–81, pp.480–496.

Rees, R., 'The Great Copper Trials', *History Today*, December 1993, pp.38–44.

Rein, M. and Schön, D.A., 'Frame-Reflective Policy Discourse', *Beleidsanalyse*, Vol.15, 1986, pp.4–18.

Roberts, J., '"A Densely Populated and Unlovely Tract": The Residential Development of Ancoats', *Manchester Region History Review*, Special Issue, Vol.VII, 1993, pp.15–26.

Roderick, G.W. and Stephens, M.D., 'Profits and Pollution: Some Problems facing the Chemical Industry in the Nineteenth Century. The Corporation of Liverpool versus James Muspratt, Alkali Manufacturer, 1838', *Industrial Archaeology*, Vol.11, 1974, pp.35–45.

Rosen, C.M., 'Differing Perceptions of the Value of Pollution Abatement across Time and Place: Balancing Doctrine in Pollution Nuisance Law, 1840–1906', *Law and History Review*, Vol.11, 1993, pp.303–81.

Rosen, C.M. and Tarr, J.A., 'The Importance of an Urban Perspective in Environmental History', *Journal of Urban History*, Vol.20, 1994, pp.299–310.

Rosen, C.M., 'Businessmen Against Pollution in Late Nineteenth Century Chicago', *Business History Review*, Vol.69, 1995, pp.351–97.

Schwela, D., 'Vergleich der nassen Deposition von Luftverunreinigungen in den Jahren um 1870 mit heutigen Belastungswerten', *Staub-Reinhalt. Luft*, Vol.43, 1983, pp.135–39.

Secord, A., 'Science in the Pub: Artisan Botanists in Early Nineteenth-century Lancashire', *History of Science*, Vol.32, 1994, pp.269–315.

Sharratt, A. and Farrar, K.R., 'Sanitation and Public Health in Nineteenth-century Manchester', *Memoirs of the Literary and Philosophical Society of Manchester*, Vol.114, 1971–2, pp.50–69.

Sheail, J., 'Green History – The Evolving Agenda', *Rural History*, Vol.4, 1993, pp.209–23.

Sheail, J., 'Town Wastes, Agricultural Sustainability and Victorian Sewage', *Urban History*, Vol.23, 1996, pp.189–210.

Sigsworth, M. and Worboys, M., 'The Public's View of Public Health in mid-Victorian Britain', *Urban History*, Vol.21, 1994, pp.237–50.

Stewart, C., 'The Battlefield: A Pictorial Review of Victorian Manchester', *Journal of the Royal Institute of British Architects*, Vol.67, 1960, pp.236–41.

Stine, J.K. and Tarr, J.A., 'At the Intersection of Histories: Technology and the Environment', *Technology and Culture*, Vol.39, 1998, pp.601–40.

Stonely, H., '"Lady" Hopwood of Middleton', *Smokeless Air*, Autumn 1952, pp.17–19.

Stradling, D. and Thorsheim, P., 'The Smoke of Great Cities: British and American Efforts to Control Air Pollution, 1860–1914', *Environmental History*, Vol.4, 1999, pp.6–31.

Szreter, S., 'The Importance of Social Intervention in Britain's Mortality Decline c.1850–1914: a Re-interpretation of the Role of Public Health', *Social History of Medicine*, Vol.1, 1988, pp.1–37.

Te Brake, W.H., 'Air Pollution and Fuel Crises in Preindustrial London, 1250–1650', *Technology and Culture*, Vol.16, 1975, pp.337–59.

Thackray, A., 'Natural Knowledge in Cultural Context: The Manchester Model', *American Historical Review*, Vol.79, 1974, pp.672–709.

Thompson, F.M.L., 'Nineteenth-century Horse Sense', *Economic History Review*, 2nd Series, Vol.XXIX, 1976, pp.60–81.

Thompson, P., 'The Building of the Year: Manchester Town Hall', *Victorian Studies*, Vol.11, 1967–8, pp.401–03.

Tomes, N., 'The Private Side of Public Health: Sanitary Science, Domestic Hygiene, and the Germ Theory, 1870–1900', *Bulletin of the History of Medicine*, Vol.64, 1990, pp.509–39.

Turner, M.J., 'Before the Manchester School: Economic Theory in Early Nineteenth-century Manchester', *History*, Vol.79, 1994, pp.216–41.

White, R., 'American Environmental History: The Development of a New Historical Field', *Pacific Historical Review*, Vol.54, 1985, pp.297–335.

Williams, M., 'The Relations of Environmental History and Historical Geography', *Journal of Historical Geography*, Vol.20, 1994, pp.3–21.

Williams, R. and Edge, D., 'The Social Shaping of Technology', *Research Policy*, Vol.25, 1996, pp.865–99.

Worster, D., 'History as Natural History: An Essay on Theory and Method', *Pacific Historical Review*, Vol.53, 1984, pp.1–19.

Worster, D., 'Transformations of the Earth: Toward an Agroecological Perspective in History', *Journal of American History*, Vol.76, 1990, pp.1087–1106.

Wyborn, T., 'Parks for the People: The Development of Public Parks in Victorian Manchester', *Manchester Region History Review*, Vol.IX, 1995, pp.2–14.

Theses

Gunn, S., 'The Manchester Middle Class, 1850–1880', unpublished PhD thesis, University of Manchester, 1992.

Richards, T., 'River Pollution Control in Industrial Lancashire, 1848–1939', unpublished PhD thesis, University of Lancaster, 1982.

Walklett, H.J., 'The Pollution of the Rivers of South-East Lancashire by Industrial Waste between c.1860 and c.1900', unpublished PhD thesis, University of Lancaster, 1993.

Index